CW00859997

AN AWESOME RUN

SELECTED WORKS, 1978-2008

LAWRENCE A. ZUMO, M.D.

authorHOUSE®

AuthorHouse™
1663 Liberty Drive, Suite 200
Bloomington, IN 47403
www.authorhouse.com
Phone: 1-800-839-8640

First published by AuthorHouse 12/1/2008

ISBN: 978-1-4389-2395-6 (e)
ISBN: 978-1-4389-2393-2 (sc)
ISBN: 978-1-4389-2394-9 (hc)

Library of Congress Control Number: 2008910919

Printed in the United States of America
Bloomington, Indiana

This book is printed on acid-free paper.

Contents

DEDICATION

TO LIBERIA'S WAR AFFECTED GENERATION, WOUNDED IN BODY OR PSYCHE, WHOSE STRIDES AND ASPIRATIONS ARE HEAVILY LADEN BUT NOT IMPOSSIBLE;

TO ALL OF LIBERIA'S CHILDREN, RICH OR POOR, WHO ARE THE TRUE BOARD OF DIRECTORS OF ANY GENUINE, LASTING POLITICAL DISPENSATION IN LIBERIA;

FINALLY, TO KESSELLY AND MALAIKA, WHO KNOW THEY HAVE TO ACHIEVE MORE THAN SUPERB GIVEN THE OPPPORTUNITIES THEY HAVE THAT THEIR PARENTS COULD NOT EVEN DREAM OF.

PREFACE:

Several decades ago while still in high school in Liberia, I really wanted to write a book about things that I thought that would be interesting to relay to other people. The whole process was still new to me .I had not gathered enough materials nor experience to make that dream a reality, so I decided to shelve the idea for some time later.

I did not know that life would get in the way and this dream would have to be deferred for three decades. During those intervening years, many things would take shape and forever change my outlook on many aspects of life. In a way those were powerful, albeit painful, wake up calls. Life with its ups-and-downs came with the full force of a gale force wind.

Thanks, goodness, I withstood those challenges and changes and amazingly I am still standing with almost the tenacity of Sindbad the Sailor, a fabled character from earlier times.In my opinion, the wait, after all, was well worth it despite all the many near misses.

Several people made this work possible. Among them: my wife , Janet, and two children, whose unending love pulled me away from that fabled cliff thus making my reach to shore so much easier. Others include Brother Edward Foken, then principal of St. Patrick's High School in Monrovia, Liberia who approved my academic scholarship at that school which enabled me to have a solid academic basis for the rest of my life; the numerous people along the way who became my surrogate parents these many years that I have been away from home since age 13; the disciplined and caring members of my extended family; my father, David Kokulo Zumo, a very strict disciplinarian and "lover of education" who instilled in me the perpetual love of learning , reading and discipline to study despite all the odds; and finally to Feribacsi and the people of Hungary who , believing in the brotherhood of humankind and wanting to speak to their own sense of this brotherhood without regard to color or national origin, warmly welcomed me and provided a very conducive intellectual environment for me to study medicine, despite it being one of the most tumultuous period in that nation's history. They took the time to welcome and nurture a total stranger from faraway land in Africa whom they barely knew. This was for me the ultimate test of the human conscience and spirit as the enormity of their national difficulties was visible everywhere during that period. For this extraordinary trust, kindness and fantastic, uninterrupted educational opportunity, I shall be forever grateful.

This book is divided into several parts. Part I- early works, which includes samples of my first real attempt at writing; Part II-high school era, which summarizes the work I did as editor in chief at St. Patrick's; Part III- post high school era, about that tumultuous period in our nation's history with all its vicissitudes; Part IV-medical school era; giving a glimpse of my work in Hungary; Part V: post training period-chronicling some of my important writings contributing to the social consciousness of our country, Liberia. This foray into sociopolitics has become necessary because the eery ghosts of Belle Yalla, Liberia's Gulag Archipelago, continues to haunt Liberia in many profound ways. In my opinion, our national history of failure, falsehood and not even knowing it runs on a parallel track to this tragic epoch of our existence. These too must be addressed and amends made or else our national journey, it seems, will be perpetually difficult. We must all grapple with lingering pains of the near depraved indifference and misguided policies of our leadership circle as well as the attendant corruption and cronyism which so mortally threaten Liberia's life, liberty and prosperity.

This, intoto, however is a brief catalogue of my journey (via snapshots of various writings) that covers four continents, spanning three decades from my impoverished roots in the jungle of Liberia, West Africa to high school editor-in-chief, then to medical school and finally as a neurologist and entrepreneur with time still left- A TRULY AWESOME RUN.

Lawrence A. Zumo, MD
Baltimore, Maryland

August 2008

PART I:

EARLY WORKS.

This short story written on September25, 1978 was my first real attempt at putting into use the English language that I had now acquired at St. Patrick's High School, Monrovia, Liberia. It is presented here in its entirety.

GOODBYE AND GOODNIGHT, LORPU (September 25, 1978)

Late Sunday night, Lorpu Flomokollie, the daughter of the rich rice merchant, Anthony Y. Flomokollie gave out a very loud cry suddenly. I woke up very surprisingly to the noise and after some minutes, before I came back to my senses I had a drink a glass of water from my makeshift refrigerator nearby. That went a long way to calm my rapidly beating heart.

As it was late in the night, many people did not wake up especially those hard working factory workers who lived in our neighbourhood.

Immediately the ambulance was called and Doctor Grimes, the doctor and ambulance attendant, was sent to come to the aid of the distressed Lorpu when she arrived at the ambulance station. Doctor Grimes had spend twenty years in the medical field and so doctoring at night was nothing strange to him at all.

Quickly he set to work and finally he tentatively diagnosed that the abscess of the neck gland which has now led to the disease of tetany. "It is characterized by violent muscular contraction and deep nervous depression", Doctor Grimes added, author, tatively. This final diagnosed was confirmed at the hospital after further testings. He concluded that it would take only a minor surgery and recuperation that would last for only a couple days and that Lorpu would be back home with her family safe and sound.

After hearing those comforting words, Mr. Flomokollie calmed down and aided the physician assistant to get his daughter into the hospital van. The Doctor packed his instrument and gave them to his assistant. Right after that the van lurched forward like a distresses lion leaping for his life.

Early Monday morning Mr. Flomokollie arrived at the hospital. By all indications, he was just about the first visitor at the hospital that morning.

The receptionist told him to wait as the doctor had not yet arrived. Luckily for me I met Mr. Flomokollie there just at the right time.

It was a lucky story for me because I had been sent by my agency,THELIBERIANUNDERCOVERSSYNDICATE, to do a cover report on the events at the hospital and since dear Lorpu was sent to this hospital, the Baker Memorial Hospital, I decided to come here and do the cover story.

Still having the deep impression of those comforting words that the doctor gave him yesterday, Mr. Flomokollie sat back comfortably in his seat as he waited for the doctor's arrival. He was as cool as a nice Christmas morning and easy as the sands of time. He sat upright looking up at the ceiling as a wandering man in search of hidden treasure but the thought of Lorpu was just like a burning sword piercing the heart of a poor soul wo had no place and part in this changing world of ours.

I walked up to Mr. Flomokollie and said "How do you do, sir?" I am in quite good health but it is only Lorpu's admittance to the hospital that still bothers me, "Mr. Flomokollie reply.

Before I knew it, Mr. Flomokollie and I had joined in a deep conversation that would not even be disturbed if a heavy rock fell on the ground about twenty yards away.

In our conversation, Mr. Flomokollie had told me how he had been so successful in the rice and art dealing business, and how he managed to keep up with his busy schedule and family duties.

He told me that he had travelled to almost all European countries, many African countries and seven Asian nations just for the quest of art masterpieces that would draw the attention of many vendors and collectors alike . He said that he was running a home business art studio that earn him an annual income of about $5 million and other foreign businesses and agents that give him the amount of $3 million per anum as royalty on the African art work that he had lend them for annual display purposes.

He told me that in his years of travels he spent close to $2 million just to cover transportation of those art items.

Upon hearing about such a great success of such a man who started out as a poor guy, I almost sent in my resignation to my agency but I just had to hold my heart and take the bull by the horn. That figure in the hands of one human being was just too great for me to imagine.

An errand boy of the receptionist came and summoned Mr. Flomokollie to the office Doctor Grimes. Immediately Mr. Flomokollie gave a sigh of relief but I could hear the sound of his heavy under-breathing.

When Mr. Flomokollie left, I followed him but instead I went to the out-patient department asking for information that would get me to see the doctor for medical attention. But that was not my intention. I wanted to see what the action of the doctor would be like. After some minutes, I asked for a medical journal and I just passed the time reading the journal while I waited for their response.

I was directed into the visitor a lounge where I was given some papers to fill in. After a further period I filled up the appointment slip that I asked for at first and put it in the middle of the medical journal.

After some minutes a boy who I suppose came from the cafeteria asked me if I wanted to have a cup of coffee. I told him "yes" and he brought me it. I answered positively because it was a rather cold morning and I needed something hot.

Glacing at my watch I knew that it had been two hours since Mr. Flomokollie left me sitting in the lounge. And I became alerted as I waited for him impatiently for an interview.

My back ached and I felt like a turtle that had stayed long enough in a shell and who wanted to come cut to ease itself. At last Mr. Flomokollie stepped out of the office of Doctor Grimes and came directly toward me just as if he knew where I would be after he left me. But something caught my attention. He seemed less comfortable and his facial expression proved that to me. So I just asked him how the case had gone and he said that the doctor was attempting a minor surgery on Lorpu, so he had to stay and see what the results would be. "As the chloroform injected into Lorpu had not gone into effect yet, I had to wait for so long a time" he concluded.

I guess, he too grew tired and decided to go home to check on a few business items. After we exchanged a few words, Mr Flomokollie hailed a taxi cab and he set out for home because he had told his driver not to come for him as he did not know the exact time he would be needing his personal driver.

I,too, set for my office so that I could file my report for the day on time. This is an excerpt of that report:

.......At 8:15 AM I met Mr. Flomokollie at the hospital's visitor lounge........... He told me that he was waiting to see Doctor Grimes........... After some time he was called by the the doctor.......... I followed him briefly but instead I went and got a journal from the receptionist. I was directed to the reading room.......... There I pretended that I was reading........ Finally he stepped out .... He told me that the doctor attempted a minor surgery but it seemed longer than usual....... So he had to wait......... He took off in a cab and I headead to my office as well due to time constraints. ----- I am to meet him tomorrow at the hospital. .... End of day's task............

The next day, Tuesday was raining but as the day progressed, the sun broke through and it became an absolute return to normal weather. I started my day with an umbrella and an over coat because I thought that the rain would continue the whole day but the sudden breakthrough of the sun rays proved me wrong. Anyway, I took a cab down the street corner and arrived at the hospital a little later.

Mr. Flomokollie had not yet arrived so I took a seat in the visitor's lounge and started out with my plans for the day. I jotted down some very important points. And I started to think about my dear friend, Lorpu who was now in the hospital for so long with no clear word about what wwas going on. I was feeling quite uneasy and impatient as we waited to see her after this supposed pre-operation stabilization and clearance. Every thing medical at Baker's Memorial Hospital always took so long and there was always no clear communication with the patients or their families and friends. There was always this sense of arrogance about all those medical people and their staff at the hospital. Even the transporters took part in this ritual as well. Although this was a government hospital funded by tax payer money, the lay public was nevertheless always cut out out of the picture with disdain. Maybe they felt that we were completely ignorant about any medical things. While many of us are not medically trained, however, they could take the time to educate some of us so we could elevate our knowledge and our level of interaction with them. After all, in the countryside where there are virtually no hospitals, the villagers take care of themselves with medicinal herbs as much as they know how without the input of these so-called medical people. But maybe this could be asking too much from those arrogant kinds. I think they just lost their human touch and they all felt they were now some kind of God to lord over us as they very well pleased. So we had to patiently wait as we always did. I tried to say a short prayer for Lorpu but I could not really concentrate on what I was saying to God, so I had to let it go for the time.

Lorpu's thought in my head made my blood to run cold. But what could I do? Nothing. I just had to sit and ask God to help my dear Lorpu.

Mr. Flomokollie arrived exatly an hour after I set foot at the hospital.

"Good morning, my man" Mr. Flomokollie said "How do you do?" "Fine Sir, it is only God that we have to thank for our health and well-being. I answered.

When did you get here, Agent Kerkula?". Well, I got here

about ten o'clock and since I did not see you, I decided to run through my plans until you arrived. And now here you are. I hope things will go well for Lorpu," I remarked.

Immediately Mr. Flomokollie filled up an appointment slip and it was sent to Doctor Grimes. But the doctor's reply was that he would be able to see Mr. Flomokollie only after he had treated an emergency that had just been reported.

"I am sure that my daughter is now longing to see me" said Mr. Flomokollie. My wife had been quite worried about the outcome of our dear beloved Lorpu. I began a bit worried last night too. So I called over the phone to the hospital but it was told that she was asleep and there was no need of disturbing her" Mr. Flomokollie continued. He said that his wife is quite pessimistic and with her conservative nature she is not that easy to convince. "This has cause a whole lot of uneasiness at home" he concluded.

Doctor Grimes had just finished attending to the case of the patient who was in urgent need of doctor's attention. The receptionist signaled Mr. Flomokollie to go and see the doctor as he was not busy at the moment.

"Good morning, Doctor Grimes, how is Lorpu?" Mr. Flomokollie began. "I have been a bit worried about the health of Lorpu but since you said that she would be alright, I just calmed down momentarily . Don't worry, she will be quite all right" said the doctor. Mr. Flomokollie asked whether Lorpu would need any toy at the moment but the doctor said that the would not be able to play with such although she was thirteen and it was those in this age group that kids respected the presence of toys very much.

He also asked if she needed any books for reading. But the doctor said that she had no time for any of such thing at the present moment.

"Tomorrow morning will be the time for the operation" said the doctor. Suddenly I could see the ripples of Mr. Flomokollie's stomach as it became upset. The surgery has been calculated to last less than one hour. He was also told that one of those young doctors from the medical school would be there on a stand-by unit for emergency cases and out-patient treatment," Doctor Grimes added.

When Mr. Flomokollie came out he briefed me about the major developments surrounding Lorpu's case and I jotted them down.

I left the hospital a little after lunch and set out for my office. I recorded the following in my file as the major report to be entered for the day. Originally, it was written in shorthand but for the benefit of all, I have recorded it here vividly in plain and simple English.

.... Mr. Flomokollie arrived at the hospital at ten o'clock, about an hour after I had arrived at the hospital, too.... He walked towards me with his fingers in his mouth and I suppose he was chewing some of his nails off........ He reported uneasiness in his feelings......... Immediately he went up to the receptionist and filled up an appointment slip in order to see Doctor Grimes......The doctor was unable to see him right away because he had to attend to an emergency case..........Finally the doctor arrived and told Mr. Flomokollie that the minor surgery will take place tomorrow, September 5, 1999........ Mr. Flomokollie showed his uneasiness but the the doctor consoled him...... Mr. Flomokollie went his way and I, my way................ (stop).........

Today is Wednesday, September 5,. It is the day to determine the fate of my dear Lorpu. It is quite calm and the stillness of air or atmosphere-whatever you may call it makes today a very nice and cool day. Many patients are here as usual but today they seen to be a bit more. By the way I have been told that this number is just right for a normal day.

Mr. Flomokollie was informed that the operation would start promptly at 8:00 am. Mr. Flomokollie was still waiting at home in order to give the doctor ample time to start the operation, knowing that 8 am prompt means something entirely different in Liberia. I guess he will stay home until after three hours and he will proceed to the hospital.

Last night he was called and told of these following proceeding so he rang me up and told me the details and offered that he would take me along in his car today.

I had my notebooks prepared. I settle down with a cup of some fine Japanese tea. Coincidently, when I took the last sip of tea, a car honked right in my front yard. And beyond all reasonable doubts, that was Mr. Flomokollie. I got in as quick as possible. And within a few moments we were on our way to the hospital.

I took a glance at my watch, and to my surprise three hours had just flown by as if it were a supersonic jet. How the time would just fly like that is one of the few things that I will ever remember about jet.

Mr. Flomokollie's driver started the engine and we dashed out like a wild horse running for its life. In less than fifteen minutes, we were at the hospital.

"Good morning, dear receptionist". Mr. Flomokollie said" I am the father of Lorpu and I am sure that you know me very well. I would like to see Doctor Grimes. ". "Yes, Mr. Flomokollie" answered the receptionist" Just give me a second and everything will be alright.

Mr Flomokollie had been told by Doctor Grimes that the operation was over but Lorpu was now in a coma.

Upon hearing this, Mr. Flomokollie became quite uneasy and even an observer who knew little about the situation would deduce that Mr. Flomokollie was at the verge of crying.

Mr. Flomokollie was that type of conservative that took every message at heart and even his dear wife Karvia, who had arrived much earlier at the hospital could not comfort him satisfactorily.

Immediately he called back home on the phone and told his other children and family members about that had happened. I too became worried. I proceeded by jotting down the message and the expressions of Mr Flomokollie's feeling in my note book.

-------Mr. Flomokollie has now gone wild about the news of Lorpu's coma------ He is now in absolute grief ------- He is at the point of loosing himself-----------Stop.........

Suddenly speculations about Lorpu's coma had gathered around the hospital and whereabouts of Doctor Grimes.

None of us present could logically deduce about the coma because we were completely ignorant of all medical theories and practices. All we could say was, how could such a minor operation lead into such a coma which is most often connected with major operation. Some doctors at Baker's Memorial were consulted but no favorable answers could be arrived at. All we could see and feel as lay people was a conspiracy of silence.

Since the time when Doctor Grimes gave the news that Lorpu was in a coma, we had not seen him.

I looked down to the gate of the hospital and I saw a car that brought Lorpu's frail grandmother, Madame Lawuo Gahnwuluku and the other grandchildren. They were directed to the visitor's lounge where we were all sitting.

She sadly moved towards Mr. Flomokollie and dropped herself to the ground in absolute agony and wailing. The two sons came around their father and mother but nothing they could do to console them. The scene, so pitiful, was too difficult to watch. We too began to shed tears for them.

After some time Mr. Flomokollie told us that he would be back. He walked straight into the hospital and he was directed into the office of the doctor.

We were all seated and we did not know what topic could break the dead silence that existed here. We all just had our eyes fixed on the hospital and the office of the doctor.

I thought my eyes deceived me but it was really true. Mr. Flomokollie came out of the hospital with full speed, shirt wide open, tie loosen, buttons all gone and he was bitterly crying. The children rushed to him and his wife embraced him. To be fair, I could not console any of them. I let my tears to flow too but no walling as the other persons present were doing.

I jotted down all these events but any tears could not permit me so I had to just close my notebook.

We asked Mr. Flomokollie and he said that Lorpu had died because the doctor said she refused to come from the coma.

Nobody could comfort the Flomokollie family. All we did was just to join the family in the time of their grief and sorrow.

The hospital police was called up and immediately they came to the surround the area where the Flomokollie family was sitting. Whether to protect the hospital, Doctor Grimes or the Flomokollie family, we will never know their true intentions.

Immediately an autopsy was ordered and the external forensic pathologist was summoned and began his work after he was fully brief about all details of the case. After four exhausting hours, the chief pathologist reported the following confidential conclusion (obtained and paraphrased in layman's term):

Lorpu's operation was a minor one but Doctor Grimes prolonged the time of the operation------ so her whole system was affected-----. Right after the operation, there was reported bleeding in her neck gland, compressing her breathing tube which cut off the air to her brain. This caused her to go into a coma------- After three hours, Lorpu died because Doctor Grimes had not done the operation in the standard and customary way. With a high degree of certainty, her death could have been avoided. Doctor Grimes is culpable on the following basis ------1)prolonged time of operation------ 2)wrong operative procedure-----3 gross medical negligence, misfeascence and malfeascence, to be finally validated by medicolegal authorities------- (STOP)

The final conclusion was that Doctor Grimes was legally responsible for Lorpu's death. All legal proceedings could then be pursued against him, once the final determination was validated by the medicolegal authorities..

Doctor Arthur Grimes was asked to join the police at the police station for his statements. I was told he casually drove there, gave a few words and returned immediately to his office, without showing any sense of remorse. I reached at my office and filed my usual daily report.

Later in the day, several Liberian newspaper headlines carried the following captions about the death of Lorpu at the hospital.

WEST AFRICAN SENTINEL:

" Doctors should be cross-examined before gaining admittance

to the medical bar and periodically monitored as long they are licensed."

THENATION'SPRIDE:NOTGUILTY,DR.GRIMES!!! IGNORAMUS! INGRATES ARE HERE TO TARNISH THE REPUTATION OF THE GOOD DOCTOR.

THE LIBERIAN SUN: Doctors who willfully kill patients should be legally executed:

THE EVENING STAR: If doctors are not liable for their malpractices, then their willful and intentional mistakes will always buried in the grave:

THE STUDENT REVIEW: If doctors begin gross malpractice, then where do we go from here?:

THE SCHOOL DISCOURSE: Why should outside vengeance be carried into the hospital and operating room by doctors?:

NOON REPORT: Let Doctor Grimes be executed today and then talk about justice tomorrow"........................

Lorpu's funeral was arranged and she was buried with full honors one sunny day in the family cemetery.

" Lord, when will fair play be ever executed in the medical profession?" was Mr. Flomokollie's last question.

"Ladies and gentleman, there the hospital saga ended or began, depending on which side of the river bank of Liberian justice you stand..

THE LOST AFRICAN IDENTITY; A PERSONAL REFLECTION

Lawrence Amos Zumo
February 7, 1980 @ 12:30 AM

Under the present circumstances, there are several forces tearing our African culture apart and making it completely obscure. I am compelled to present a few views concerning these destructive forces.

With the theme being a very general one, I will present my views from the Liberian situation — the one that immediately affect those locally termed "country people", the true advocates of the African culture and those not regarded in any .

It is a shame that those who speak about the unfair treatments given to our African culture, our unique representation, are the very ones that are the architects of these destructive strategies directly or indirectly, no matter how innocent they may proclaim to be. In all aspects, each and every one of us are greatly responsible for the destruction of our unique and distinct African culture and traditions. We can say that we are all culpable to varying degrees on this issue.

To those Liberians who term their fellow human from the hinterland and other villages in Liberia "country", I say woe unto you because you have defeated the very purpose for which you were created and has brought shame on your folks and kinsmen and have shown that you are completely ignorant of the fact of what a true Liberian is.

We have too many times regarded others as inferiors because they are not civilized or westernized as you claim to be. If we were in our right minds and not influenced by outside forces then these remarks probably would never have been made. These assumptions stem from the fact that we are ignorant and really don't understand what we ourselves are saying to others. If we were only willing to only love those "inferiors" as we call them, we would have greatly benefitted ourselves for they are the ones who are the true masters of the African culture and traditions.

To those who have been termed the "inferiors" because of their so-called uncivilization or unwesternization, I commend you for keeping alive that ever-burning torch of African image. We also are aware that you are being mentally and socially tortured by these handful few because they are extremely envious of your heritage. I salute you and admire your courage. Stemming from the fact that they (the name callers) are unable to faithfully live the true African life as you have done, they show their envy for these that they are unwilling to accept because thy have been brainwashed. They call you provocative names, socially categorizing you and financially isolating you as signs of their unending envy. Keep trekking on!!

With all these, you will always be remembered as the true heroes and heroines of the real Africa. If fustrations have made you to change that true path which you followed, I admonish you to pursue the former for the rewards are great and inmeasurable .

Do your part to pass on the torch of this true African

culture to your children and their children's children, by teaching them your languages and unique cultural ways.

To those who have abandon the African culture think twice. You may be making a great mistake. The African culture is too rich to be casted away as you have done in the past. Look in your hearts and minds and see whether this is the right thing you are doing.

These beautiful and rich African dialects which characterize our Africanness are the same ones that we are wiping away willfully and replacing them by those "unsatisfactory" languages from faraway lands.

You should know by now how difficult it is to express your true feelings in those languages. We have been made to believe that they hold the key to the true elements of human civilization. Is that really true? You be the judge.

Wishing you good luck in your endeavours, let me remind all that we should be the master of our own destiny from where we stand, not from another man's land.

PART II:

HIGH SCHOOL ERA:

As editor in chief , I was in charge of the originality, content and clarity of the school newspaper, which we published about four or five times during the senior year. Given the success of our efforts, I was asked to be editor of the final newspaper issue of the graduating class. In that issue, a sample of whose contents appear below, I tried to sum up the most important events and thoughts of the key players of our class. I also attempted to include parting words , thoughts and reflections as well as futuristic projections of the destiny of my fellow classmates at that time. At that time, I had hoped that this would be our contribution to Liberia's literary repertoire as well as point of reference for years to come. The result of that attempt follows below:

THOUGHTS FROM THE PRINCIPAL

BRO. DONALD ALLEN, C.S.C.

I wish to thank the Senior Class for this opportunity of presenting your graduation. It is quite difficult for me to filter out one or two of the many thoughts that run through my mind at graduation time. One thought I want to stress is what I will call "WISHES". What do I wish for you ? Certainly, all of us at St. Pat's and other staff members, and students too

wish you all much success and happiness in your future life, in whatever direction it may take you. Some of you will be going on to university studies. Others will seek employment and become self-supporting. Some will leave the Monrovia area and a few will leave Liberia in order to continue their education in another country. What is certain in the midst of this uncertainty is that you will never be together

again in such large numbers. Will this, as it often does, result in the loosening of your bonds of friendships with each other? Hopefully the good times and happy experiences you have shared will be a strong link to keep you close to each other. So,

another strong wish I send you is that of continuing your unity with one another. May the petty irritations and vexations of the school days be forgotten and give place to a genuine appreciation of the gifts and goodness that lie in all of you. In the future may you continue to be as close in spirit as you are now geographically. My hope and wish is that you take an active part in whatever capacity you can in the St. Pat's Alumni Association. I hope that a few of you keep the others informed of what each is doing even though you are separated by a good many miles and also by interests and occupations.

Another wish: May you never lose the sense of gratitude for what your parents have done for you. Especially may never lose the sense of gratitude you owe God, who created you, gave you blessings and brought you thus far. Good luck and many blessing to you all in the exiting future—awaits you. This, in the name of the teachers, students and on my own----

OUR CLASS HISTORY :

Lawrence A. Zumo & Philip A. Matthew.

Our adventurous and most interesting voyage started on the 3rd of March, 1975. There were 81 courageous , highly intelligent and ambitious young men who started this remarkable voyage. Because we were so many, we decided to divide into two groups. The first group was composed of 42 daring young men under the leadership of John Massaquoi. The other group was composed of 39 brave young men under the leadership of Mr. Louis Kamara. With these arrangements, we started our voyage at 8:00 on Monday morning on March 3, 1975.

Both ships travelled together on the voyage for the period of nine month without much difficulty but at the latter part of the tenth month of our voyage both ships were attacked by violent storms, which brought us a great loss of many of our best

friends. The 7-1 section topped the year's honor roll list and became the 1975 table tennis champs as well during their voyage.

The following year, 1976, we discovered that out of the original 81 young men, we had only 48 left. Some of our friends had left us because they could not stand the hardships which we had to pass through, especially those violent storms which came once every six weeks. Since we had been reduced from 81 to 48, we decided to sail as one group, on board one ship under the leadership of Mr. Lawrence A. Zumo. On a Monday morning in February 1976 we were joined by five energetic young men: James Benjamin, Ervin Cooper, Jervis Witherspoon, Nimene Kun, and Joseph Tolbert thus increasing our number to 53. After sailing for ten months again, most of our friends left us because they were unable to sail through the bad weather, and as a result we were further reduced to 38.

After being anchored for about three months, we continued our voyage in February of 1977 under the leadership of Mr Louis Kamara . We were again joined by four young ambitious men. Fianyo Gbedemah, Robert Marcy, Patrick Toe, and Steven Witteeven and this increased our number to 42. Together we started the

difficult voyage again. When we reached our next destination, we discovered that two of our old members of the group had also left. Eventhough we were greatly reduced by then, we were very fortunate to have been joined by some very intelligent, ambitious men. These gentlemen were: Henry Cooper, Gwyn Seyon,

Everett Townsend, Khalil Huballa, Magnus Krakue, Charles Lincoln, Standford Peabody, and Samuel Vansiea. We were also glad to have Mr.Eldred Nims, who left us in1978, rejoin us. This, then increasing our number to 50. This voyage of 1978 was one of the very challenging ones and it was made under the leadership of

Theophilus Toe. After traveling for ten more months of hardships, most our friends decided to leave because they found it difficult to continue the voyage with us. We then eagerly anchored to rest for three months.

After this brief rest, we decided to continue under the

leadership of Augustine Jarrett during the latest part of February 1979. That Monday morning, we gladly had Mr. James T. Nimley rejoin us. Mr. Nimley had earlier left us in 1976. We sailed again for ten months of violent storms and many others difficulties. At this time, we had been reduced by many hardships to only 32 at the end of the ten months of hard struggle.

We were very glad to have Mr. Judson Neal as our leader for the last part of the voyage. The last stage began on February 18, 1980. This was the most remarkable period of the voyage, as at this time we were familiar with each other and shared some nice jokes and fun in class.. Eventhough at this stage of the voyage many of us were weak because of the hardships we had passed through, we tried our possible best to reach our destination. At the end of the voyage, we found out that out of the 81 men who had originally started the voyage, only 19 left to end the voyage with the new members who came to join the original group. This was how our most adventurous, hard and interesting voyage took its course.

HISTORY OF ST. PATRICK'S HIGH

By Lawrence A Zumo

The aim of St. Patrick's High has always been to turn out Christian gentlemen of the highest calibre who can take their places with distinction and credit in all walks of life.

St. Patrick's School, comprised of St. Patrick's Elementary

School situated on Snapper Hill and the present high school here on Capitol Hill , had its beginning in very hallowed quarters that is, the sacristy at the Father's old residence in the year 1921.

Among the pioneer fathers of St. Patrick's high, the following were outstanding: Fathers McEnry, Baker, McAndrew, O'Leary, Coleman and McHew. In 1922, the present edifice of the Sacred Heart Cathedral was constructed and

classes began in the church. Later, the senior classes were held upstairs on the veranda of the Father's residence. The year 1934 saw the erection of the first school name engraved on the edifice as " St. Patrick's School. Principals of the school during the period following this opening included Fathers R.J. Duffee, and Peter Rogers. In 1939, Bishop

Francis Carrol (then Father), who had already been principal for two years, remained in this capacity until he turned out after a period of four years, another group of graduates in 1943. These graduates in addition to satisfying the school curricula, qualified on the Junior and Senior Cambridge University Standard. It would be quite in place to mention here that it was the intention of Monsignor

Carroll to plan and to train students for the Cambridge Junior and Senior Examinations. He flew to Freetown where he talked with the fathers in the Catholic Schools there. He brought back the curricula used in the Sierra Leone schools. After he discovered that this would not work out satisfactorily, he decided to raise the school to high school level.

The standard of the founding curriculum as outlined or as stated earlier continued well beyond 1950 until the government decreed thal all schools should operate on the same standard curriculum.

This meant the adoption of the standard of the government

schools. However, the St. Patrick's High School standard continued to be much higher than average.

In 1947, Father Joseph Guinan became principal of the school. Next to Father (late Bishop) Carroll, he saw the completion of the new wing to the old school building here on Capitol Hill. To Father Guinan goes the credit of reviving the Boy

Scouts, Cadet Corps, and making St. Patrick's team into the champion of the Inter-School football teams.

After negociations, the brothers of Holy Cross arrived in the middle of 1962 to assume the responsability of administering St. Patrick's High School, which up to

this time had been under the supervision of the Irish priests of the society of African Missions (SMA). The party of brothers was led by the late Brother Theophine. Among others, the party of brothers included Brother Donald, who was the first brother of Holy Cross to become principal of St. Patrick's high School. The era of the brothers has brought a total change both academically and physically. The curriculum has been rearranged and intensified so as to meet the standards of modern high school students. The students were introduced to and encouraged to use the library. The laboratory was modernized and in sports., basketball was introduced. The School did not only witness an increase of the number graduated, but also an uplifting of the calibre of the students.

In the year 1963, the beloved Bro. Theophine Schmidt, CSC met a sudden death which resulted from an automobile accident in Liberia. He was succeeded as superior by Bro. Francisco. Brother Francisco served in this capacity until the end of 1965. He introduced art in St. Patrick's High School. Later, Bro. Donald was relieved of his position as principal and was succeeded by Bro. Austin Maley, CSC in 1964. This year now saw an increase in the number of brothers.

By giving awards at the end of each year and introducing the honnor roll system, the Brothers have inspired the students to show excellent results. When the "ZOES" (class of 80) arrived at St. Patrick's in 1975, Brother Edward Foken was then the principal. He left us before we completed our most interesting journey at that school.

An attempt at poetic expression of thanks to St. Patrick's for my solid high school education:

DEAR ST. PAT'S
To thee, o dear and beloved St. Pats
Do we owe all our achievements and
successes; with God above and with
our parents; to thank for their great
care and love; we promise to live our
lives with sincerity.

Oh, how sweet a memory to think of the
years; that we spent together at St.
Pats; within its four walls to prepare
for a brighter future
So sweet, that "twill be the last to
forget

Dear brothers and faculty members, to
you always; no matter whenever we go
"Hats off"

though life here was not always bright
As God did not promise joy without
tears; we thank you for correcting
our mistakes; though was a
painstaking job, we thank you

O great and dear Liberia, to you we
come; To give sincerely what St. Pats
has given us
In our own weak ways, here's how
we can donate our quotas to society

Parents, Hats off to you, teachers,
hats off to you
With God above, we promise sincerely
To do our parts to uplift our SOCIETY.

THE LAST WILLS AND TESTAMENTS(CLASS OF "80)

Lawrence A. Zumo(ed)

I, James Edward Benjamin, hereby solemnly will and bequeath my seat in the senoir class to Adele Adighibe, my set of books to Robert Cooper, provided he is willing to pay $25.00 for them, my height to Eric Ricks, my position as guard on the school team to Alston Wolo, my handsome appearance to Lloyd Diggs, my mechanical drawing ability to Eli Saint-Pe, my well disciplined behavior to Thomas Dundas, and last but not the last, I will my love to faculty, staff, and the entire student body of St. Patrick's High School.

I, Amos Aristotle M. Boyd, do hereby will my seat in the senior class to Sie Collins, my athletic abilities to Alben Tarty, and my strokes to Christopher Buckle.

I, Joel Carter, will my intelligence and unique leadership abilities to Lloyd Diggs, my seat to Alvin Jones, my ability to pass literature to Joseph Armer, my non-smoking habit to Christopher Buckle, and my ability to get along with others to Thomas Dundes.

I, T. Benedict Giple, do hereby will my seat in the senior class to Sampson Wuor, my creditable ability to do mechanical drawing to Robert Cooper, my height to Kofi Woods, my position in the cadet Corps to Joseph Collins, and a bit of my knowledge to Rufus Berry.

I, Finayo Gbedemah Jr., will my eating ability to Mr. Massaquoi, our school's cook, my good manners to Sylvester Thomas, my study habits to Chris Stevens, ans lastly, my love to the entire body of St. Patrick's

I, Samuel David Glover Jr., do hereby will my books to Velmer Porte, my seat in the senior class to Mana Sherman, my good grades throughout the year to Wolo Tweh, and my high level on intelligence also goes to Elmer Porte.

I, Khalil Huballa will my senior class chair and position in the school to Khaled Abu Rujaib.

I, Ronald Hoff, will my Jersey (no.10) on the senior Varsity

Basketball Team to Alston Wolo, my seat in the senior class to Mana Sherman, my good conduct to Adrian Hoff, and my height to Kofi Woods.

I, Gus Jarrett, do hereby solemnly will and bequeath my seat in the senior class to my former classmate Joseph Armer, who decided that he would graduate in 1981, my 3200 test booklet to anyone in class 11 who will pay for it, my excellent English ability to Lloyd Diggs, my drafting ability to Eli Saint-Pe, my size to Eric Ricks, my books to Thomas Dundas, my sobriety to Sylvester Toe, and to the rest of St. Patrick's family, I will my love.

I, Lawrence Amos Zumo, do hereby will and bequeath my journalistic ability to Sylvester Toe, my mathematical ability to Thomas Toh, my ability to understand others to Ben Koryon, and my easy going attitude to Emmanuel Paygar. I also do solemnly will my curiosity to learn to Augustine Flomo, my determination to James Gibson, and finally, my love of principles to Abraham Borbor and to the whole St. Patrick's family.

I, Charles D. Konuwa, president of the Student Council of St. Patrick's for 1980, do hereby will my political ambition to Chris Stevens, my reading ability to Wilsin Saytarkon, my seat in the senior class to Joseph Dennis, my ability to get along with others to Thomas Toh, and lastly my journalistic ability to Sylvester Toe.

I, Nimene Kun, hereby will my intellectual ability to Lawrence Kun, my extra sensory preception to Wilson Saytarkon, and my journalistic ability to Sylvester Toe.

I, Magnus B. Krakue, hereby, without any skepticism, pronounce my will which is to be effective from this date. I will my extraordinary abilities in sports, academic curriculum, technical and scientific know-how, and my humanistic concern to Alvin Jones. I, in light of the above, I sincerely hope that you Mr. Jones, will utilize these unique abilities to the best of your capabilities.

I, Charles Lincoln Jr., do hereby will my leadership ability as captain of the Varsity Football Team to Alben Tarty, my cooperative attitude to Joseph Collins, my seat in the senior class to James Walker, and lastly, my duty to keep the classes locked during the lunch period to Timothy Kie.

I, S. Adino Philip Matthew, being of sound and disposing mind, will my books and all others class materials to Ben Koryon, my official uniform to Emmanuel Paygar.and my daily uniform to James Gibson. I also will my hardworking habit and good conduct on campus to Augustine Flomo. Lastly my eating habit in the cafeteria to James Flomo.

I, Jallah Mensah, will my good times at St. Patrick's to Issac Watson. I also will my plate, spoon, and cup to him.

I, Eldred Nims, hereby will my seat in the senior class to Mana Sherman, my standing in the class to Joseph Armer, my intelligence to Sylvester Toe, and finally my singing ability to Merrill Badio.

I, Solomon W. R. Patray Jr., do hereby will and bequeath my mathematical ability to Timothy Wulah, my table tennis ability to Raffee Wright, my seat to Lloyd Diggs, and my military ability to Othello Ezeagu.

I, Stanford Peabody, sane in the mind, do hereby will my football ability to Michael Taplah, my super coolness to Marcus Jones, my seat in the senior class to Edwin Harmon, as long as he wishes to stay there, my versatility in soccer, basketball, swimming, and all other time consumers to my brother, Alec Peabody. These abilities he is to have without any strings attached, until he is of status and class to will them on to anyone else he sees fit.

I, Charles Saygbah, will my sense of generosity to James Kpadeh, my cooperative attitude to Chris Stevens, my sense of justice to Kofi Woods, my taciturnity to Kwee Fahnbulleh, and lastly, I will my easy-going attitude to Samuel Corvah.

I, Rizvon Shakir, will my mathematical intelligence to Naresh Harjani, my most respected class behavior to Lal Mahtani, my handsome pesonality to Ajit Matthews, and my commanding ability to Mr. Mensah of class 11.

I, Nyenawreh Toe, do hereby will my desk to my brother Sylvester Toe, my basketball ability to Lawrence Wollor, my football dexterity to Osborne Hill, my greater participatory abilities in school activities to Wilson Saytarkon, my coolness at school to Wolo Tweh, my hard studying ability to Philip Tarr, and lastly my strategic dancing technique to Jonathan Frank.

I, Joseph J. Tolbert, of sound mind and memory, do hereby solomnly will my seat in the senior class to Leo Marshall, my set of books to Edwin Harmon. My maturity to Thomans Dundas, my studying habit to all students in the school that need it, my creative thinking ability to Robert Cooper , my position on the varsity basketball team to Lloyd Diggs, and last of all, I would like to express my love to the school and all the teachers that contributed to the success of my education at St. Patrick's.

I, Everett Bismarck Townsend, with a clear mind and memory, do hereby solemnly will and bequeath my seat and standard in the senior class to George Satiah. My complete set of books to Mana Sherman, and my good artistic ability to Adele Adighibe. I also will my excellent behavior on campus to Sylvester Thomas, my good sportsmanship to Paul Jarvon, and finally, my appreciation and thanks to the faculty and staff of St. Patricks.

I,Gwen Thomas Seyon, of sound mind and body do solemnly will and bequeath everything I have to anybody who doesn't have. And on behalf of my family I would like to extend my love and thanks to the entire family of St. Patricks for what they have done for me.

I, Samuel Vansiea, do hereby will my intellectual ability, love and commitment to Christ, to my beloved brother Nathaniel Flomo Vansiea, my curiosity to learn to my dear friend Peter Wureh, and to my friend Benedict Walker, I will my honesty, my truthfulness and my hardworking ability.

I, Boima Voyou, will my football ability to Kwee Fahnbulleh, and my good behavior to Lloyd Diggs.

I, Henry Cooper , in a clear state of mind, do hereby ask the most divine power to pass on my greatest appreciation and regard to Bro. Thomas, for showing me that you don't have to take everything personal. Also my total encouragement to the basketball and soccer team members and coaches. Last but not the least, my thanks and bio-chemical good -bye to Mr. Henrique Smith.

I, James Theophilus Nimley Jr. will and bequeath my intelligence to Sylvester Toe, my bombastic eloquence from brother Edward Dailey to the class of 1981 and my creative ability to fulfill greater philanthropic goals to Joseph Sie Collins.

I, J. Blamoh Kunwon, do hereby will my outspokeness to Peter Koffa, my captainship and ability to lead to Thomas Dundas, my seat to Joseph Armer, my ability to kiss to Michael Taplah, my position as last Lt. Colonel to Lloyd Diggs, my dividing ability to Benjamin Koryon, my radioactivity to Mr. Alex Anderson, and finally my singing ability to Brother Donald Allen.

I, James Judson Neal, being on good health and of a sound mind, do hereby will and bequeath my position on the Cadet Corps (colonel) to Merrill Badio, my seat to Kofi Woods, my academic abilities to Christopher Stevens, Sylvester Toe, and John Smith. I also will my position on the varsity soccer team and my #4 outfit to Paul Jarvan, my soccer techniques to Michael Taplan, my scientific abilities to Rufus Berry. My mathematical ingenuity to Jeremiah Nabwe. To Othello Ezeagu goes my position as president of Grade 12. Last but not the least, to Eric Ricks, my philathelic enthusiasm.

I, Ervin Cooper, being a man of strong courage and patience and intelligence hereby will my strong courage and patience to Wolo Tweh, urging him to use this as a sword to purge forward in life.

WHO'S WHO ON CLASS 12 TEACHING STAFF.

Lawrence A. Zumo (ed)

Br. Donald J. Allen: Br. Donald Allen has been the Principal of St. Patrick's High School for a little over two years. He was born on March22,1926 in the U.S. He obtained a Bachelor's and Master's degree in Zoology at Michigan State University . Later, Br. Donald Allen received a M.A. in Religious Education from Boston College. He is one of those principals who is concerned about raising the standard and respect of the school.

Mr. Alexander Anderson: Mr. Anderson, who is the Dean of discipline of St. Patrick's High School, was born on August 29, 1949. Mr. Anderson is commonly known in the school as "PSYCHO" After completing high school, he did much of his higher education on his own. He has been at St. Patrick's for the past five years and he is our economics teacher.

Br. Edward Dailey, CSC: On November 5, 1938, Br. Edward was born in the United States and is one the most highly qualified teachers of our school. He received his B.A. and M.A. degrees in 1960 and 1961 respectively from the University of Notre Dame. Later, he acquired his Postmaster's Diploma from Michigan State University in 1969. He has been at St. Patrick's High School for 1 ½ years and he was our French instructor.

Br. Thomas Dillman, CSC: Born on the dawn of the Great Depression, namely, August 28, 1927, in the United States. Br. Thomas has led a varied life for the past 53 years. He has taught Physics and Mathematics at St. Patrick's High School for the past twelve years. Apart from other qualifications which he refused to mention, Br. Thomas has had thirty years of teaching experience.

Mr. Kofi Kwarteng-Amanin: Mr. Kofi, who is from Ghana, was born on August 15, 1954. After his completion on secondary school, Mr. Kofi proceeded to the University of Ghana, where he obtained a B.A. (Honours) degree in Sociology with Economics in 1978. He taught history at St. Patrick's High School.

Fr. Louis W. Rink: Fr. Lou Rink, born on February 13, 1924 in the United States. Apart from acquiring degrees in Philosophy and Theology, Fr. Lou's wide experience ranges from the U.S. Army, to nearly 21 years experience in teaching. He had taught fourteen years in Uganda before proceeding to Liberia in 1974 .He taught us English and Doctrine at St. Patrick's High School.

CLASS PROPHECY

Lawrence A. Zumo & Augustine Jarrett(contributors)

A few nights ago, I was out in town with my friend. We had a few bottles of beer and Guiness stout. At about twelve midnight, we decided to go home. I dropped him at his house, I was too tired to stay up and eat so I went to bed.

.... All of a sudden a light with eminent brilliance shone and many things that my eyes have never beheld were revealed to me. Call it a vision if you like, but I prefer not to name it so. Anyhow for now to avoid too many ifs , I shall refer to it as a "revelation".

I entered the future 2000 and I was walking down a street, which appeared to be the same place where we now have Randall street. As I walked down this street, I came to a marvel of architecture which was oddly beautiful. There was a sign which read " the Carter Building". I entered through a laser beam operated door. While descending the escalator, I saw a person whom I recognized as Joel Carter. There were two men walking with him.

I asked him about our other classmates but he said that he was on his way to confer with a geological company. I wondered what that had to do with our classmates.

We arrived at this place and after waiting for some time, we were ushered into the office of Dr. Ben Giple. We discussed how I could provide this venture with insurance and banking needs.

I left Carter and Giple discussing and went back to my office. By the way, I was also President of the West African Insurance Firm.

At about ten minutes before lunch my intercom buzzed and announced that a Mr. Krakue and a Mr. Tolbert wanted to meet with me. I recognized them as my old friends. They were opening a company that manufactured planes and wanted a comprehensive insurance coverage on this billion dollar project. I told them I would need more information so they decided to another meeting. I asked if he had seen "Stumpy" he said no, but added that the last thing he heard of him was that he had resigned his post as executive vice president of the National Bank of Anteagua, and has turned to a life of shooting pool and drinking beer.

I had the chance of visiting the J.F.K. Medical Center and the doctor I met was the Chief Medical Officer, Dr. Lawrence Amos Zumo, a very skilled urologist. We talked for a long time in his richly carpeted air-conditioned office. He told me that he had a company and that other brains like Dr. Sam Vansiea and Samuel Glover were major share holders.

Some Lebanese business men came to me in Liberia as they were sent by an old friend. Khalil Huballa who was the head of a pharmaceutical company in this country. They confirmed that Mr. Nimene Kun had qualified as a parasitologist and was working full time in Lebanon. These specialist wanted a building in Liberia so I contacted the construction Firm of Cooper and Associates to draw the plans and do the work. Although Henry was the owner of this company. He really didn't need the contract the Lebanese were paying him for.

John Blamoh had advanced from a Captain to the rank of commanding general of the Armed Forces of Liberia. The first four star general in many years.

A small boy with the features of an Indian came to a white bearded man with a letter. On this letter was the address of Dr. Rizvon Shakir who, as I learned later, had become an Electrical Engineer and a Computer scientist with Ph.D's in both.

This was not the last revelation of the success of my fellow classmates . Charles Lincoln was now a millionaire after successfully completing his studies in banking and finance.

News reached me also that a Perry Mayson of our times had entered the Liberian legal profession. Finally I realized that he was no one other than Charles Saygbah.

Solomon Patray was now one of Liberia's best bone specialist and had a successful practice in central Monrovia.

James Judson Neal had a good Job and had received fame from the medical profession as being one of Liberia's foremost surgeons

The leading agronomist at the Liberian Institute of Soil Science and Research located in Robertsport was Dr. Charles D. Konowa who had returned from the U.S. some years ago.

I was also told by the revealed that the development of roads and highways, and the construction of houses had reached its peak. These were all engineered by Everett Towsend, one of our best civil engineers.

I was told also that Liberia was now ahead in industrial development and technological advancements. These were all made possible by the ingenious works of Theophilus Toe, one of the few best Liberian geologists.

Mr. Jallah Mensah was the governor of the National Bank of Liberia and had made the mark of excellence. Philip Matthew had become a successful mechanical engineer and had established a plant that assembled cars.

Amos Boyd was a psychiatrist who had mastered his profession so well that people did not mind the fee of $20,000.00 per month he charged for his excellent and professional service.

Ervin Cooper was the owner and president of his own Finance & Trust Company.

Boima Voyou had become a very successful medical doctor. A general practitioner. He was always so busy that no one could ever remember when he was out of his white medical suit.

Stanford was also a medical doctor and was deciding to go to England to specialize.

Although Mr. Gbedemah had returned from the U.S. after specializing in his area of interest, he was made the Ghanaian Ambassador to the United Nations.

Mr. James T. Nimley, had become a scientist, the first Liberian ever to pursue that area of interest.

Ronald Hoff, Eldred Nims, and Gwyn Seyon were now prominent Liberians in high positions of trust.

That's how the revelation went and I was very happy after all that my former classmates had become quite successful in various walks of life.

The climax of my academic experience at St. Patrick's High School was when I was proclaimed the valedictorian of the graduating class of 1980 and had to follow in the tradition of those before me who had to give the famous Valedictory Speech at the E.J. Roye Building in Monrovia, Liberia.

VALEDICTORY ADDRESS: DECEMBER 7, 1980

BY LAWRENCE AMOS ZUMO

Your Grace Bishop Michael Francis
Our distinguished Guest Speaker,
Principals & Members of the Faculties of
St. Theresa's & St. Patrick's High School,
Parents & Guardians,
Members of the graduating classes,
Visiting guest, Friends, Well-wishers,
Ladies & Gentlemen:

Today certainly begins a new chapter in our lives. For some of us, the attainment to this level has come from parental emulation; for others, it is one of the many steps demanded by their familial ethos; yet for others, it is a marked improvement over the academic achievement of their parents. Whatever our category today, it is most significant to realize that a certain degree of of progress has been made. As for me, I have been and will ever be mindful of the great opportunity that has been afforded me. To this end, I have pondered this question; IS SETTING PRIORITIES ESSENTIAL TO OUR FUTURE ADVANCEMENT AND/OR DEVELOPMENT?

Before I briefly attempt to answer this question, let me define three words. According to the Webster College Dictionary, PRIORITY is " that which needs or merits attention before others". Randon House Dictionary of the English Language defines "ADVANCEMENT as " the act of moving forward" and DEVELOPMENT as the act or process of bringing out the capabilities or possibilities".

As I perceive things, in order for one to progress successfully there must be a plan of action. This presupposes an organized attitude toward fulfilling some determined objectives. Consequently to move forward, in advance, one must fully exploit his capabilities; i.e develop. To do this, I believe one must seek the highest possible education made available to him/her. In our age of technology and rapidly advancing scientific discoveries, qualified expertise is not anymore a luxury. Clearly then in Liberia the need to develop along these lines must be made exigent. Often times I get the impression from discussion with my peers that if they had to choose between a difficult course or an easy one, they'll opt for the easier course. The explanation is " one does not have to subject himself/herself to an obviously difficult situation when there is an alternatively easy choice". Be that it may, it would seem to me that these individuals have set their priority-take the course of least resistance, least work, least challenge.

I am not advocating that one's priority must by its very nature involve hardship. The point I wish to make is that it must involve SACRIFICE . What seems difficult sometimes appears to be because we do not want to forego our complacency or sometimes our comfort. Let me illustrate. An individual with an above average I.Q. opts to take a degree in physical Education (no offense to people in this area) than say computer science. It is not because this individual does not have the ability to do computer science, but rather he prefers to "hang out" every evening. Clearly computer training and "hanging out" do not mix. If he has the stamina (and I know a lot who do), he can certainly "hang out" and exercise the next day. Admittedly, the situation is not as simplistic as this. The point I wish to bring out is every career area is as important as another. No matter how we play one against the other, the fact remains that a good physical educator is just as necessary as a good "anything". It serves one no purpose if he/she compromises standard. If one chooses to do physical education, I would have thought that he would sacrifice during his preparatory training to do well and become competent. If one decides " to cut corners" or " to bluff his way through",. he/she usually ends up wanting, deficient and ultimately frustrated, leaning more towards envy and corruption.

Consequently, another element to setting one's priority is truth. We have to "be true" to ourselves. We have to do our utmost in whatever we do. We have to live within the framework of choices, never deluding ourselves but always attempting to respond within our limitations. It doesn't help us to be oblivious to our limitation; to involve ourselves in activities or escapades which spell disaster right from the word "go" to conform because it is the "in thing" to do, to get by because we indulge in other important things. Clearly, Shakespeare's admonishment to be true to oneself implies the weighing of all alternatives in the light of what obtains for each individual and then proceeding. Sometimes, the choice is hard and may seem pointless, but so is life.

It would seem that I am giving the impression that one can achieve the setting of priority exclusively and entirely on his own. Nothing can be further from the truth. Jesus Christ has made us the promise that in Him we have abundant life. He urges us to take" His yoke upon(our) shoulders and learn from Him"......... For (His) yoke is easy and (His) burden light.". He has also asserted that He is the "way, the truth and the life". Ultimaltely he is the way in our lives. Following his precepts and his demands are a sure way to setting our priorities right. At times, we shall come to the realization that we are entirely capable of determination of our own destiny. We shall think that our educational achievements solely provide us with the credentials that enable us to do things on our own. However, in these situation that we become anxious, begin to wander, or become disillusioned, let us be mindful that Christ's invitation to "remain on Him", relying totally on His Grace and guidance is a claim that works.

Fellow graduates, in looking to the future, let us always be cognizant of the roads we'll face. At times the barriers may be insurmountable but in Christ we can overcome them. Centrally put, Christ in us sets the priorities of our lives. This may seen unpopular in our area, but believe me it works.

I would like at this juncture to say thanks to St. Patrick's, the brothers, faculty and staff, and fellow schoolmates. The road has been long and arduous . At times, one surely could have given up. The advice, encouragement, camaraderie, and support exhibited along the years clearly are responsible for this event today. How can we say farewell? I shall only say we leave you because this a natural progression in the sequence of our live's events. However, our fond memories of "dear old St. Patrick's" will forever endure.

Before closing, I wish to seize this moment to honourably thank those who helped me inorder to be able to graduate from St. Patrick's. I wish sincerely to acknowledge the fact that without the help of those many individuals I would not have be able to reach thus far. I would like to thank the Brothers of the

Holy Cross who negotiated with the government to get me a Scholarship inorder to fund my stay here at St. Patrick's. Many persons aided me in getting through high school. However, it will take me hours to list all their names here. Therefore I wish to tell them all "THANK YOU". But before concluding that, I wish to thank Mrs. Evangeline Bracewell for the extreme generosity shown me during the course of this year.

Last but not the least, my parents, Mr. David K Zumo & Mae. Paykue, who suffered a lot for my well-being and who corrected my many mistakes along the way. When I reflect on their many efforts and suffering just for my sake, my heart mourns. However, I thank them for making me the person I am today. I wish you all God's blessings in your many endeavours and hope that your insistence and God's grace will lead to the determination of right PRIORITIES in my life. I THANK YOU !!!

PART III:

POST-HIGH SCHOOL PERIOD:

The military coup of April 12, 1980 in Liberia was one of uttermost confusion for many Liberians, let alone young high school students like us, because of myriad of mixed feelings about that event and the anxiety about what the future would hold for all.

Below are a few thoughts expressed poetically about events pre and post this event. I will also include some essays about events before that fateful day as the political temperature was no longer at 32 degrees Fahreinheit since the April 14, 1979 rice demonstration and its aftermath. We began to look deeper at who were as humans and as a nation in relation to other classmates, students who were at different strata of Liberian society. The new path of introspection was very anxiety-and-ambivalence- provoking for many years to come.

12th April, 1980- A Day to Remember

12th April 1980 vividly proved
That time and tide change so fast
That we are no match to them both
Even if blissfully ignorant we are.

When in the dead of night
Sound of shells awoke us
In this glorious land of liberty
Tommorow would be so,so far away

A purpose, a genuine one
That rattled this glorious land of liberty
Turning it into a calamity
As 14th April 1979 has foreshadowed

After 133 years of elite hegemony
Mayhem, slavery, injustice intoto;
Our brothers in the barracks with tears;
Sleep their new courage did not deter.

Unknown to civilians and foreigners alike
Concrete plans for upheveal were underway;
In the dead of that fateful Friday night
Seventeen young men went gallantly to end servitude.

Their lives to God they gave that night;
Had they failed on that sacrificial bridge
Unimaginable, indescrible the ensuing nightmare
That would have followed.

In the immediate post high school period, during my several encounters with different nationalities at that time, I had a particular exchange with a university lecturer. At the end of that exchange, I was given a book by Colin Turnbull about the Ike of Uganda. I was asked to compare and contrast their situation with what I saw in Liberia. That essay included below.

A Comparison Essay (1981): Ik of Uganda and the natives of Liberia

The story on the Ik of the Karimojong region of Uganda written by Mr. Colin M. Turnbull is a very sad portrayal of an originally very ambitious and strong people-hunters. In his story (his book) on the Ik, there are some instances where clarity needs to be added-eg. the author wrote that if the others knew that Lomejas's son Ajurokingomoi had died the day before, they would expect him to give a feast. What I don't understand is how could they expect that, knowing that there was severe famine in the land and they were supposedly a very individualistic people. I also find it very difficult to readily absorb some of his finding on the Ik people but no means doubting person but rather may even laugh at that person. This is something that I have never experienced before in Africa because in Africa no matter what the situation the people go at lengths to do whatever possible to save a dying person. Finding amusement in someone else's death in Africa has been something that I never dreamed of.

Comparatively, in Liberia when the Americo-Liberians (former plantation slaves who set sail for Africa in the 1820's to found the republic of Liberia) arrived there, they became very stern with and inhumane to the natives whom they met. They (Americo-Liberians) forced them to give up their traditional practices, their rights in communal land ownership, and their traditional beliefs inorder to "civilize" them. The manner in which this was presented was totally unacceptable to the natives and thus they refused to accept those measures. The Americo-Liberians in anger overpowered them and forced them into subjugation and degradation. The natives were faced with no other alternatives but to accept the Americo-Liberian rule.

After a period of time, not being able to further withstand the forces of the Americo-Liberian politics, the natives were forced to retreat into high equatorial forests of Liberia. However, when they retreated they did not forget those virtues-kindness, goodness, generosity. etc. That they felt were the basics of their society although it was a struggle of the survival of the fittest.

They did not become a loveless people. They cared for each other. The sick and the dying were treated with the use of all available traditional herbs. People shared their cath with others. The older people and the children were taken care of. The dead ones were honourably buried and the sick ones were never laughed at. Everybody showed utmost concern for each other and kindness was the watchword of their society. There were however a few of those natives who preferred to be deviants. Those were made to conform by guidelines set by others.

It is however in some cities today in Liberia that those things the author observed among the Ik people are now occurring but at an unnoticeably lower level. The present economic system with some of the problems that is has imposed is suggesting the people become more individualistic than ever before.

Morever, I agree with the author on the grounds that

we very often take for granted th that we practice in our own individualistically-oriented societies. He mentioned that the individualism that we practice may one day reach to the level at which it is now practiced in the ancien society.

Very often, we are guilty of putting our drives ahead of our values and that practice in itself is imposing an overbearing danger which we are not willing to readily see and accept.

Who knows when? Time may be running out or not but sooner or later, we may be faced with accepting the ugly and painful realities of our individualism. I may be wrong, I may be right. The truth is that I myself am not quite sure when that could happen. Maybe society will devise means to solve those problems in the future, I hope.

THREE YEARS IN A STUDENT MOVEMENT

By Lawrence Amos Zumo (1981)

The three years that I spent working in a student movement in Liberia gave me first hand experience that I could not have experienced otherwise. Among them the government crackdown on student leaders, the inner organization of a student movement, motives of major student actions, the plight of the workers, and student actions in the classrooms.

In 1978, I joined the YCS, a Christian-based student movement affiliated with the LYCS in Paris, France and International Union of Students in Austria. I joined the movement out of the mere curiosity of knowing how it worked, but after three years of active membership I am proud to say that I have no regrets.

When I joined the movement, our first major task was to make a critique of action at the secondary school level. This mean that we had to observe some of the problems secondary school students encountered and how they solved them.

At a particular secondary school in Monrovia, St. Theresa's Convent School, we observed that some members of the senior class constantly disturbed the class lectures by talking out of order and by misbehaving. These disturbances sometimes angered the instructors and the lectures were often stopped until the students left the class.

One day, we were invited to visit this school. Upon accepting the invitation, our five member team swiftly began to analyze the problems. After three days of hard work, among other things, we found that all students in the senior class were not in favour of these disturbances. The non-trouble makers were helpless because they did not have the power to take any appropriate actions. As a result, they suffered because the class time was always taken to stop the disturbances ans thus the course outlines were never fully exhausted. In our final analysis, we urged the non-trouble makers to take the matter to higher school authorities if the disturbances should continue in the future.

When in 1979, events on the political scene predicted some radical changes that would affect the lives of everyone, we made it our duty to closely study these events and analyze the motives behind the frequent student actions. We, therefore, read all release that came from the Student Union headquarters at the University if Liberia and every newspaper article that we could find. We also asked people's opinions on some of the unconstitutional legislations that our government was making i.e. giving management the sole power to dismiss any worker at its will and pleasure, making any unions formed by any financed-agencies illegal, and the banning of all strike actions by workers in the private sector.

After months of collecting data of these events, we were asked by the International YCS to present this report at an international student conference in Sierra Leone, another West African country. In our report, we observed that our government, was making important decisions and powerful legislative acts that excluded the bulk majority of the population. Among other things, we observed that the public officials used public finance at their will because they were not answerable to anybody else except themselves.

Unfortunately, that report brought us many misfortunes. When we returned to Liberia after the conference, some government officials managed to get a copy of the report and we therefore met stiff opposition from the government. In less than two months, at the peak of the national crises, two prominent leaders of our organization along with student leaders from other youth organizations around the country were imprisoned on very trivial and baseless charges. I was, therefore, left with no other alternatives but to become passive until the crack down was over.

LIBERIA'S DEVELOPMENTAL STRIDES: ANOTHER LOOK- July 1986

Throught the years (especially the past ten years), I have had the immense opprtunity of experiencing various aspect of life. This gave me the opprtunity to change some views and has also aided me to be better aware of my environment.

Many theories and subjections have been offered by Liberians and non-Liberians alike to explain the tremendous economic problems that are -plaquing Liberia and the other nations in the third and fourth worlds respectively. Although the literatures of economic development abound in these theories etc, it is still not quite clear why the conditions of poor nations continue to worsen while wealth abound elsewhere. Political leaders must consider some of the suggestions. I will attempt to address some Liberians issues that may have been too often underestimated, misrepresented, or scantily discussed.

The historical events leading to the underdevelopment of Liberia, and those leading to the 1980 military coup d'etat should be quite clear by now. Historians write that in the 18th and 19th centuries, the inhabitants of what is now Liberia were exporters of agricultural produce and other elements of commerce. The transition from communalism to the present individualism and marked dependence on imports is frankly far from clear. Research in the areas as well as many others should be pursued by historians, sociologist, educators, scientists, etc. and the results should be made public for the public good.

Four major concerns that often pop up when one considers the Liberian millieu are a) attitude adjustments, b) infatuation with artificial desires, c) the problem of ideologies, and d) the quality of education offered at the primary, secondary, and tertiary levels. As our policy makers are afraid of drastic changes, our attitudes towards these changes and towards various aspects of sociatal issues require critical analyses. Without such analyses and possible implementations, material developmental strides will prove futile. While it is true that one's attitudes, perceptions, and goals are shaped very much by the environment , as thinking being, one can also reciprocate equally with your surroundings. Our heads deserve better use and this one of those ways.

The conflicts between the Americo-Liberian and the indigenious Liberians, the successful oppression of audacity among the Liberian populace, the collision of ideas between western civilization and traditional African values, and the blind acceptance of all artifacts of western culture have left us all in a difficult spot. This requires urgent investigation by all concerned for the future. This certainly has a great role to our present attitude problems and the inhumane"segregation" classification of our society today..

To many people, being civilized is understood in different ways. That is all and good, once you don't fall over the cliff with wrong ideas about being civilized. The unparalleted technological advance in the west are proceeding at an unprecendents pace. We have to realize this, put aside little talk and get on with what we have to do- for our destinies are beckoning. Africa missed the Industrial Revolution of the 18th Century, will we miss the new technology revolution of the 21st century?

As a consequence of our historical past, we have not been able to adequately deal with several problems that have come our way. A few examples are mentioned below: Liberians will prefer to buy many goods made in the U.S.A. as opposed to those identical goods that sometimes are produced locally. I am still puzzled about why many Liberians will prefer to buy for example a used motor-bike, being sold at an exhorbitant rate from a Lebanese trader ,than buying a new, cheaper motor-bike of the same make and model from a Liberian trader. Many Liberians at home and in America, for example, have enslaved themselves to American and European fashion trends. They see the search for a sound education, at best , a burden,at worse, an obstacle. Such beliefs are utterly,unintelligent, stupid and with out any concrete justifications. On another front, if a Liberian citizen is denieda visa to travel to the United State of America, he comtemplates suicide at all. Several decades ago, Western propaganda chiefs successfully glamorized their countries as the land of opportunity, advancement, freedom and "honey". And we obediently accepted these. These effects continue to decimate Liberia and its people. When will we reverse this trend??????

PART IV:
MEDICAL SCHOOL PERIOD

After the lingering chaos of the 1980 coup, the revenge, counter revenge life became the more uncertain. Liberians for the first time began to think of places other than Liberia and the United States of America to go to school for further advancements.

With so much soul searching and complex decision making, gamble and significant sacrifices, I ended up at the University Medical School, Debrecen, Hungary in September 1986 on a wing and a prayer, literally. The next six years there would be one of the most difficult financially that I ever encountered. Somehow I pulled through with flying colours on a very, very strong academic background after a fabulous medical education although I was financially challenged throughout.

Besides the basic medical education, I was interested in extracurricular medical scientific research. Here is a result of that effort. A published research paper in the journal: Biochimia et Biophysica Acta in 1989 summarized below. Full text available as noted. During the research leading to this publication, I was glad to work alongside and learn immensely from Hungarian academician, Prof. Pal Gergely and superbrain, Distinguished Professor Pal Kertai. An excerpt of that same paper was presented at the

27th Congress of the Student Scientific Society (Kaunas, Lithuania) of the Baltic Republics in November 1989. I received a certificate of excellence from that Scientific Society for that presentation on behalf of the Hungarian delegation that I was leading at that time.

After my return to Hungary from that conference, I was appointed in the Medical School Institute of Human Physiology as Laboratory Demonstrator from 1990 to 1992- a position I served with distinction. Interestingly, it was in this same institute decades earlier that Hans (Janos) Selye did the seminal research work to describe and codify today what we know as " stress". I was delighted to work in the shadow of that scientific giant and others who came before and after him.

Purification and partial characterization of protein phosphatases from rat thymus.

Bakó E, Dombrádi V, Erdödi F, Zumo Lawrence, Kertai P, Gergely P.

Department of Medical Chemistry, University School of Medicine, Debrecen Hungary.

Protein phosphatases assayed with phosphorylase alpha are present in the soluble and particulate fractions of rat thymocytes. Phosphorylase phosphatase activity in the cytosol fraction was resolved by heparin-Sepharose chromatography into type-1 and type-2A enzymes. Similarities between thymocyte and muscle or liver protein phosphatase-1 included preferential dephosphorylation of the beta subunit of phosphorylase kinase, inhibition by inhibitor-2 and retention by heparin-Sepharose. Similarities between thymocyte and muscle or liver protein phosphatase-2A included specificity for the alpha subunit of phosphorylase kinase, insensitivity to the action of inhibitor-2, lack of retention by heparin-Sepharose and stimulation by polycationic macromolecules such as polybrene, protamine and histone H1. Protein phosphatase-1 from the cytosol fraction of thymocytes had an apparent molecular mass of 120 kDa as determined by gel filtration. The phosphatase-2A separated from the cytosol of thymocytes may correspond to phosphatase-2A0, since it was completely inactive (latent) in the absence of polycation and had activity only in the presence of polycations. The apparent molecular mass of phosphatase-2A0 from thymocytes was 240 kDa as determined by gel filtration. The catalytic subunit of thymocyte type-1 protein phosphatase was purified with heparin-Sepharose chromatography followed by gel filtration and fast protein liquid chromatography on Mono Q column. The purified type-1 catalytic subunit exhibited a specific activity of 8.2 U/mg and consisted of a single protein of 35 kDa as judged by SDS-gel electrophoresis. The catalytic subunit of type-2A phosphatase from thymocytes appearing in the heparin-Sepharose flow-through fraction was further purified on protamine-Sepharose, followed by gel filtration. The specific activity of the type-2A catalytic subunit was 2.1 U/mg and consisted of a major protein of 34.5 kDa, as revealed by SDS-gel electrophoresis. (Biochim Biophys Acta. 1989 Oct 9;1013(3):300-5).

As part of the university requirement for all graduating medical students (called "szigorlo" in Hungarian), we had to work on, write and defend successfully a supervised MD Thesis, in addition to clinical work, state board examination, written and oral final university exams. I was fortunate to complete my thesis under a demanding but fair clinical supervisor. Below is the full text of that MD thesis, for posterity.

THYMECTOMY IN MYASTENIA GRAVIS; CASE STUDY OF 55 PATIENTS-August 1993

Medical University of Debrecen
Surgical Clinic No. 2

THYMECTOMY IN THE TREATMENT OF MYASTHENIA GRAVIS :A CASE STUDY OF 55 PATIENTS

Submitted by : Lawrence A. Zumo, MS(VI)
Trainee Intern
University Medical school
Debrecen, Hungary

Thesis supervisor: Sandor Kollar, MD
Assistant lecturer
Adjunktus
2nd Dept. of General

Table of Figures

Acknowledgement

A Thousand and One Thanks

To Prof. Arpad Peterffy, head of the 2nd surgical department, for permission to do the work for my MD Thesis in his institute; to Prof. Lukacs Geza for my first true surgical experience; to Dr. Kollar Sandor for his supervision and guidance; to Mae and Gbeymah for making life, in this present form, possible; to the numerous others, unnamed due to space constraints, who in one way or another helped to make this episode, and indeed this project, a tangible reality; and finally to the Almighty for providing inspiration, perseverance and an intellectual environment sufficiently conducive in which to work, despite all the odds.

1. 2. INTRODUCTION

In this thesis, we present the results of our experience with thymectomy in 55 myasthenic patients, over a 10-year period (1981-1990), who were referred to our institute for surgical intervention. 40 (72.7%) were females, whilst 15 (27.3%) were males. The average age was 30.

Myasthenia gravis, a relatively uncommon nervous system disease, occurs at an estimated prevalence rate of 0. 5-5 per 100,000 population and an incidence rate of 0.4 per 100,000 population (i.e. prevalency = 5-50 per 1 million population; average being 30 per million) (1). The disease is commonest in women (3:1-4:1) in the first four decades of life; the incidence thereafter essentially being equal (2).

Most of our patient (60 %) were operated within 1 year of diagnosis. Our data indicate a favourable remission rate post-thymectomy (38.5 % versus the literature value of 30 %). We attribute this most significantly to operation being performed at an early stage of the disease. Graphs and histograms are included to show different aspects of our case study.

3. HISTORICAL PERSPECTIVE

Myasthenia gravis (MG, Erb-Goldflam disease), a rare neuro-muscular disorder with its pathologic fatigue and rapid exhaustion of the striated musculature under the voluntary control of the nervous system, has intermittently sparked the interest of clinicians and researchers alike since its first description.

At the turn of the century the clinical picture was described phenomenologically. In the 1930s, when the chemical nature of the neuromuscular block found to be the basis of the disease was recognized, interest in MG again resurfaced (3). The singular observation by Dr. Mary Walker in 1934 of how a patient under her care resembled a case of curare poisoning and the dramatic results shown when she administered a curare antidote, eserine (physostigmine), to that patient did much to establish MG as a treatable clinical entity. This discovery was made at the same time when Dale and Feldberg were establishing the role of acetylcholine as a transmitter substance in the sympathetic ganglia of cats. It was suggested that in myasthenia gravis too rapid destruction by cholinesterase of the acetylcholine liberated at neuromuscular junctions was responsible for the weakness. Although this is now known to be incorrect, treatment with anticholinesterase drugs stabilizes the acetylcholine and allows contraction of affected muscles to occur without undue fatigue (4).

An interest in MG, continuing even at present, once again resurged when in the 1960s, the immune nature of myasthenia was demonstrated. A few reasons could be postulated for this resurgence of interest. For example, as a model disease, numerous basic questions of neurophysiology and pathology can be studied. At present it is the only neurologic disease with a proven autoimmune pathomechanism, for which both the target organ as well as the antigen and antibody are known. It is one of those severe disorders that can be treated successfully and even cured (5).

Thomas Willis, a monk-physician from Oxford, is considered the first person to describe MG. In his work, DE ANIMA BRUTORUM, published in Oxford in 1672, he gave a movingly accurate description of the clinical symptoms of the disease. He also recognized and described the diurnal fluctuation (-pathognomonically the most important -) of the symptoms.

Two centuries later Samuel Wilks, a physician at Guy's Hospital (London) in 1877, observed a 22-year –old girl, whose disease, characterized by double vision and general weakness, had previously been treated as hysteria. (She later died in hospital due to respiratory crisis).

In 1879 Erb described the condition and course of the disease of three of his patients. He described diplopia, disturbance of speech and swallowing and recognized the occurrence of sudden death from the disease, as well as the possibility of remission in one of his patients, (-that-which he attributed to his electric treatment -) (5, 6).

In summarizing the symptomatic picture and course of MG based on the observation of three patients, Goldflam in 1893 stated definitely that this disease differs fundamentally from the known types of bulbar disorder on the account of lack of atrophy, fasciculation and electric degeneration reaction. Jolly, who in 1895 coined the term myasthenia gravis paralytica, described a useful diagnostic process in which decrement of twitching was observed in muscular tension or provoked by indirect electric current. He also described the phenomenon that prolonged tiring of some muscle area may also produce fatigue in muscles apparently unaffected by MG. Some decades later Walker (1938) demonstrated this phenomenon before the British Royal Society and it has ever since been known as the Walker effect (5, 6).

Figure 1 Surface Anatomy of the Thymus (See Refernce No. 6)

4. SURGICAL TECHNIQUE

We used the median sternotomy approach. This gave us a total access to the anterior mediastinum and also an adequate exposure of the lower part of the neck.

A summary of the operative procedure is as follows (7): A standard midline incision is performed. Diathermy is used to divide the periosteum between the attachment of the pectoral muscles to the sternum and the upper end of the sternum is displayed with blunt dissection. Curved forceps are passed around the edges of the sternum on each side, keeping close to the bone to avoid injuring the internal mammary vessels. The sternum is divided with a power saw. A self-retaining retractor is inserted and the superior mediastinal contents are displayed. The layer of fascia covering the thymus is identified and divided in the midline. The plane of separation between the two lobes of the thymus is identified by blunt dissection. Dissection proceeds inferiorly, freeing the inferior poles of the gland from the pericardium and then following the lateral borders on each side. The thymus in distinguished from the mediastinal fat tissue by the difference in texture. When both lower poles are mobilized, they are elevated with tissue forceps and dissection proceeds carefully behind the gland until the left brachiocephalic (innominate) vein is exposed. The upper poles of the gland are freed by blunt dissection as far as possible, and finally released by gentle traction.

After complete hemostasis has been insured, the wound is checked to see if the pleura has been opened or not. Small hoes in the pericardium can be sutured and the wound is closed with a narrow bore suction drain to the anterior mediastinum. The divided sternum is approximated with interrupted wire sutures. The pectoral muscles are sutured together in the midline and the subcutaneous tissues and the skin ate closed (7).

5. OTHER SURGICAL APPROACHES (REF. 10)

5.1. Transcervical Thymectomy

This approach is reportedly used for routine thymectomy in myasthenia gravis in several centers. It is said to be easier in children and young adults with an elastic thoracic cage. It has an undisputed advantage of a painless postoperative course and an aesthetic scar but bears the danger of hemorrhage, pneumothorax and the risk of leaving thymic remnants.

5.2. Extended Thymectomy

This associates extracapsular resection with anterior mediastinal fat exenteration from neck to diaphragm and hilum to hilum. This technique probably contributes to the excellent remission rate, about 60 %, in non-thymomatous myasthenia gravis and currently better results in thymomatous myasthenia gravis are obtained than with other methods. Recurrence cannot be totally excluded as thymic tissue can exist outside the mediastinum.

5.3. Combined Thymectomy

When the neck exposure is considered insufficient through a presternal incision, it is possible to combine it with a transverse cervicotomy. Jaretzki defends this approach arguing that the gland has many cervical and mediastinal variations.

5.4. Median Sternotomy T-Incision

The skin incision is T-shaped, the horizontal limb overlying the second costal cartilage. The T-incision has the advantages of median sternotomy with good exposure of the mediastinum, safe resection of the thymus and a gratifying cosmetic aspect.

5.5. Transverse Submammary Thoracotomy

A horizontal incision is made at the upper border of the sixth rib. The incision is upwardly curved at the extremities. Sternotomy is performed vertically. The technique must be performed scrupulously as risk of postoperative hematoma, infection and skin flap necrosis exist. It also leaves a small area of insensibility between the breasts in nearly 50 % of patients.

5.6. Horizontal Thoracotomy and Sternotomy

A transverse skin incision 15-20 cm long, is performed at the sternomanubrial junction. The sternum is transected in the shape of an inverted V. This approach is purported to make possible a "thorough exploration" of the mediastinum.

5.7. Anterolateral Thoracotomy

A high anterolateral thoracotomy for a non-enlarged gland is used by only a few authors. The incision, which gives an excellent view of the mediastinum, is hidden in the submammary fold. However, the exploration of the cranial poles of the thymus seems to be difficult by this approach (10).

6. Therapeutic modalities

The treatment of MG is fairly diverse, often complicated task. It can be stated that methods of treatment have developed to such an extent that a considerable number of patients can be cured. In other instances, it is possible for the patient to lead life at the highly improved, nearly normal level and to return to creative activity.

It can be safely said that there are only few neurologic diseases that can be treated as successfully as MG. It must be stressed, however, that only 30-35% of patients can be treated adequately and in due time cured. These patients can, therefore, live without symptoms nor complaints nor having to take drugs and they can continue their original occupation; lead normal lives in every respect with no, or only rare and brief recurrences.

The following general principles have now been recognized for the timing and sequence of therapy in MG (8):

a) anticholinesterase drugs are useful in all clinical

forms of the disease, particularly, they are the mainstay in ocular MG;

b) immunosuppresants increase the frequency of remission, but have some detrimental side effects;

c) plasma exchange has only transient effects and does not confer greater long term protection than do immunosuppresants alone.

Listed below are different therapeutic intervention in MG as dictated by the circumstances and expressions of the disease:

CHOLINESTERASE INHIBITORS: Pyridostigmine (Mestinon) and less frequently neostigmine (prostigmine) are prescribed. (regular dosage is 4x 0.5 tabs per day; 1 tab = 120 mg; duration of treatment dictated by the symptoms, etc.). Spironolactone may serve as an adjuvant.

CORTICOSTEROIDS: Prednisone may be given in high doses from the start of the treatment. Initially the symptoms are frequently aggravated, therefore, treatment should always be started with continuously available supervision and respiratory aid. The noted clinical improvement induced by prednisone correlated well with a decrease in antibody titer. This effect usually occurs over 3-6 months.

IMMUNOSUPPRESSIVES: 6-Mercaptopurine or azathioprine are used in elderly subjects and also as support therapy in patients showing as unsatisfactory response to corticosteroids.

PLASMAPHERESIS: Recommended for patients in whom he disease is worsening or in cases being prepared for thymectomy. It is also used as a therapeutic umbrella for exchange transfusions to combat myasthenic reaction of the neonate in myasthenic mothers.

THYMECTOMY: Thymectomy is today, with specific reservations, the treatment of choice. The results are particularly impressive in women under age 40 who have suffered from the disease for less than 2 years. In such patients, thymectomy convincingly improves the prognosis, so surgical exploration is always indicated. The prognosis in non-neoplastic thymic lesions is the better, the more the germinal centers in the extirpated gland.

In patients over age 60, no macroscopic evidence of thymic tissue exists and no germinal centers are present; therefore on a theoretical basis, no effect can be expected from thymectomy in the elderly. The effect is better if thymectomy is followed by prednisone treatment. The operative mortality in a large case series amounted to only about 3 %. Operation should be offered to patients with roentgenologically visible thymic hyperplasia or thymomas, as well as to patients in whom no abnormal thymic tissue can be detected. Due to technical dangers and the risk of leaving thymic rests behind, the technique of partial extirpation during medianoscopy has been abandoned in favour of conventional sternotomy for thymectomy (9).

7. Pathology

The pathologic alterations in MG appear in the striated musculature, the motor end-plate and thymus (11). There are no demonstrable pathologic findings in the central nervous system. Possible damage to the peripheral nervous cannot be verified either, except in cases where MG is associated with nuclear paralysis, such as eg. Concurrence of MG and amyotrophic lateral sclerosis (12). Three stages of the alteration in the autoimmune-damaged striated muscles have been observed: coagulation necrosis, round cell infiltration (lymphorrhea) and fiber degeneration with inflammation.

The nature of the damage is varied: the incidence of neurogenic and myogenic (dystrophic) damage is the most frequent, while fluctuation in fiber caliber, fragmented fibers, proliferation and central location of the nuclei are also frequent. Hyaline (Zenker) degeneration and proliferation of the connective tissue may occur, as well as round-cell infiltration into the perimysium or in the form of lymphorrhea (lymphorrhea differs essentially from the active, round cell infiltration in that it respects neither the endo- nor the perimysium, not even the muscle fibers. In as much as it covers the structural elements, it may be compared to necrosis). The presence, nature and extent of muscular damage show no correlation with the severity and clinical characteristics of the disease nor the areas affected but they do so with the time elapsed since the onset of the disease and the result of thymectomy. A definite connection, however, has been found between the severity of the ocular symptoms, which are hard to influence, and the extend of damage to the musculature of the eye (on the basis of the biopsy material obtained in the course of operations to correct ptosis) (13).

While light microscopic examination is sufficient for routing and diagnosis, additional information may be obtained by ultrastructural studies. The alterations can be of different degrees and heterogeneity. Even in the same muscle, several degrees can be observed –from the almost healthy muscle structure through increase in connective tissue to sever fiber atrophy.

The basic pathologic phenomenon in MG is the partial or complete blockage of impulse transmission. Dystrophic and dysplastic alterations of the motor end plate could, in some cases, be demonstrated by light microscopy.

Additionally the presynaptic membrane may be small, shrunken, and the postsynaptic membrane elongated; the number of junctional folds decreased; and their surfaces fragmented and shrunken. The regular junctional folds may be broken up and made irregular by secondary and even tertiary synaptic gaps and folds.

Persistence, hyperplasia or tumor of the thymus constitutes one of the oldest and best known findings in the pathology of MG. in spite of this, the pathogenic role of the thymus is not known with certainty. Some kind of activity of the thymus was found in 70 – 90 % in various large patient operation series. According to the literature, thymoma occurs in 10 – 18 % of the cases. In our practice, thymoma accounted for 12.7 % of the cases. The activity of the thymus in MG may be measured by determining the number of germinative centers 0-1, 1-2 and 3 or more germinative centers per low power microscopic visual field correspond respectively to degrees 1, II, and III of thymic hyperplasia; while if there are no germinative centers but a lymphatic tissue characteristic of the thymus, it is termed persistent thymus.

Activation of the thymus from its inactive state may occur following an inflammation of the upper respiratory tract occurring near the thymus. This hypothesis is in line with the practical finding that the disease often starts (in some 15 % of cases) or is exarcerbated (30 % of the cases) during of following respiratory tract catarrh. This theory concerning thymitis also contributed, using the analogy of autoimmune thyroiditis, to MG being regarded as an autoimmune disorder (14-15).

In MG, the proportion of T and B lymphocytes of the thymus changes; the outer walls of the germinative centers are composed of T lymphocytes, whereas, interiorly, there is an increase in the number of B lymphocytes. It has been observed that the antibody-dependent cellular cytotoxicity (ADCC) of peripheral blood lymphocytes decreases after active treatment affecting the thymus and that the mitotic activity of the thymic lymphocytes change in response to immunosuppressive treatment. Concerning the onset and especially the autoimmune mechanism of MG, there are ancient myoid cells or cell particles in the thymus raising the possibility of intrathymic antibody production against other muscle constituents. Tolerance may cease and an autoimmune process may start against the ACHR (acetylcholine receptor found on the surface of muscle cells) or against other muscle constituents. Due to the termination of immunological tolerance, antibodies against ACHR are produced intrathymically and the immune reaction may ensue.

It also seems plausible to postulate that mutations of proliferating thymic cells produce "forbidden clones" which lead to the production of autoantibodies, in disregard of the immunotolerance to endogenous tissues. These mutating clones lead to a situation in which an immunologic reaction is provoked in the thymus, as if they were directed against a foreign tissue.

On the basis of the antigen association between the thymus, thyroid and muscle tissues, the autoimmune reaction thus provoked may be directed against the last two tissues as well, eventually leading to myasthenia gravis (15). Regarding the aforementioned hypotheses, only indirect evidence and guarded opinions can be found in the literature.

MG associated with thymoma is a separate subset within the subject of thymus pathology. It is known that MG associated with thymoma is resistant to drug treatment, disease progresses faster, therapeutic reactibility poor and the otherwise characteristic symptoms fluctuation less marked. According to the type of epithelial cells the tumors can be characterized as dark-cell or light –cell epitheliomas. MG is most frequently associated with epitheliomas with minimal lymphoid reaction. Although the clinical course and prognosis of MG in not correlated with the structure of the thymoma, the light-cell epitheliomas were clinically malignant; whilst the dark-cell thymomas tended rather to be benign as were the tumors of predominantly lymphatic structure (26).

Young women with myasthenia commonly show the HLA B-8. Dw3- Drw3 haplotype. The prevalence of MG is approximately 3 per 100,000 population. Before age 40, women are more commonly affected; MG in the elderly affects mostly males. Familial cases are rare, but other autoimmune diseases are found in the relatives of MG patients (16, 27).

Despite all these data that largely point in one direction, it has not been clarified how the anti-ACHR antibodies cause mysthenic muscle weakness or fatigue.

Presumably, several mechanisms play a part. As seen, the antibodies bind to the ACHRs on the junctional folds of the muscle membrane, thus exercising a direct blocking effect on the receptor zone of the end plate.

A further possible mechanism can be derived from the fact that besides IgG, binding of C3, C4 and C9 can also be found on the postsynaptic membrane, thus complement dependant lysis may develop in the ACHR zone with the decrease in the number of receptors on the one hand; with the development of characteristic morphological alterations (elongations, dysplasia, dystrophy, fold fragmentation and expansion of the synaptic gap) on the other. Finally accelerated degradation of the ACHRs seems to be confirmed in MG, associated with a disordered slowing down of resynthesis such that the normal turnover, rhythm of ACHR decomposition and resynthesis is disturbed. Recently, the possible role of antibodies other than those against ACHR in the pathomechanism of MG has again been suggested. Yamamoto (1986) confirmed the role of antiflamin and antivinculin antibodies in MG and polymyositis, as opposed to progressive muscular dystrophy (17).

LEGEND (AND EXPLANATIONS) TO FIGURES

Figure 1 : females (n = 40, 72.7 %)
Ratio: (2.7: 1)
Males (n = 15, 27.3 %)

Figure 2: Osserman classification (reference 3)

Stage 1: local, non-progressive MG, which Is mostly ocular, at times affecting only one eye. (it is often resistant but prognosis is generally good.)

Stage 2: Generalized MG, which affects several (II.a) groups of muscles, (tend to go into remission, reacts well to drug treatment and has good prognosis)

Stage 2b: acute MG, with fulminant bulbar symptoms, with the possibility of early respiratory crisis. (it does no respond well to drug treatment, with a generally poor prognosis.)

Stage 3: late severe MG, developing in general two years after the form belonging to groups 1 and 2, with a poor prognosis.

Stage 4: MG characterized by muscular atrophy which develops rather quickly (within 6 months) after the generalized disease and has a poor prognosis.

Figure 3:Period between establishing diagnosis and
thymectomy

Figure 4:Histologic typing

Figure 5:Postoperative results: (Remission = no sign
Of disease; Asymptomatic =no disease while taking up to 120 mg of neostigmine daily; Improved=positive changes as compared to Preop. State; unchanged; worse; died.)
(based on reference 24)

In the 10- year period of 1981-90 there were 55 myasthenic patients treated with thymectomy.

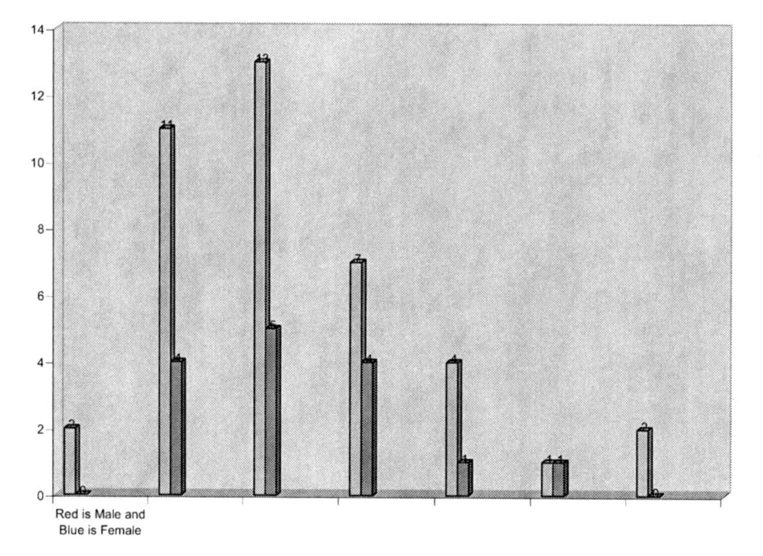

The diagnosis and preoperative stage were based on clinical classification (Osserrman)

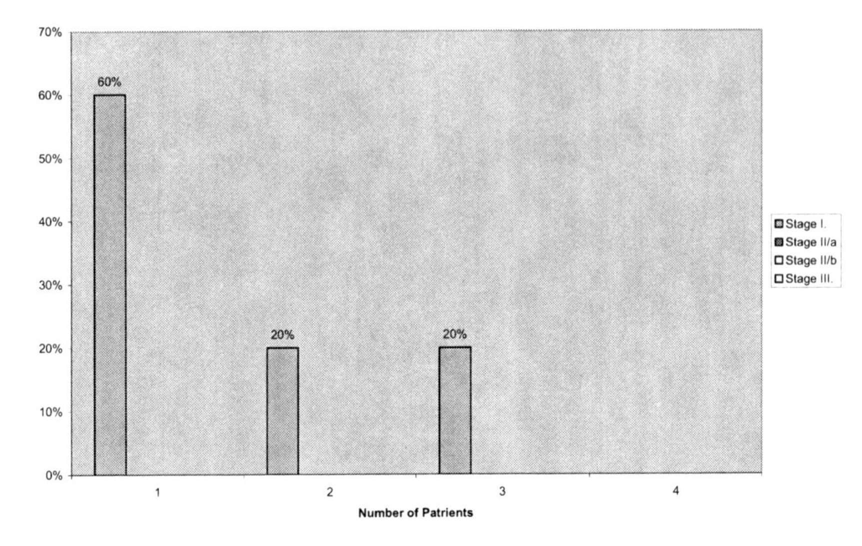

Period between establishing diagnosis and thymectomy

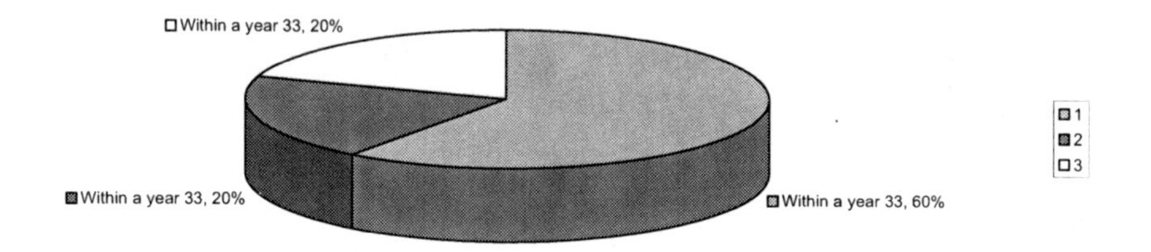

Histologic types:

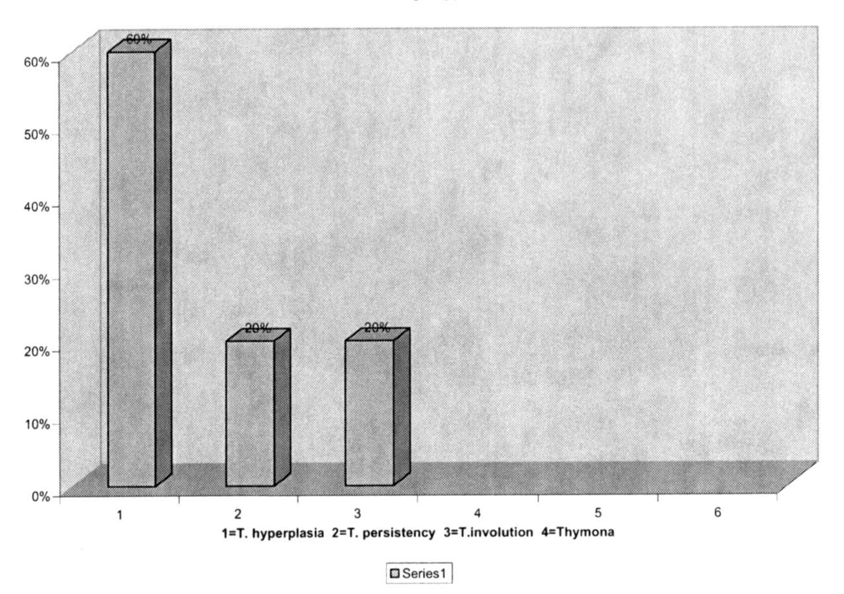

Postoperative results

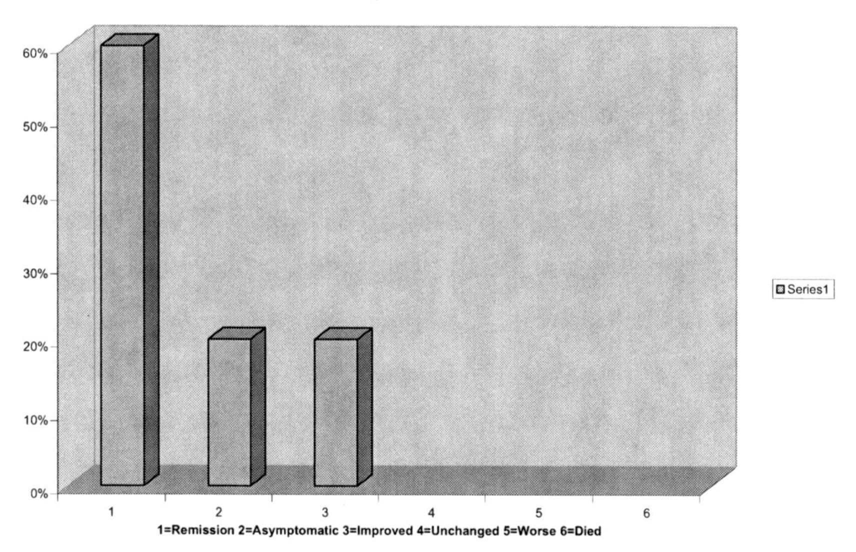

MATERIALS & METHODS

In the 10-year period, 1981-90, 55 myasthenic patients were

referred to our institute for surgical intervention. The bulk of our patients were referred to us by the Debrecen University Medical School's institute of Neurology & Psychiatry and the Kenezy County Hospital. Of the total number, 40 (72.7 %) were females, whilst 15 (27.3 %) were males. The average age was 30 years. From the histogram of patient distribution in figure 1, it is evident that the largest number of our patients were in the age group 21-30 (13 females, 5 males); followed by the age group 11-20 (11 females, 4 males).

To eliminate the possibility of leaving thymic tissue behind,as often occurred with eg.the transcervical route, we favoured and used the complete medianostomy (transternal) approach. Thymic tissue was generally situated beneath a layer of areola tissue. It was readily identified and removed. (Refer to surgical technique section.)

RESULTS AND DISCUSSION

The diagnosis and preoperative staging were based on the Osserman clinical classification (fig. 2).

4 patients (7.3 %) were in stage 1; 23 patients (41.8 %) in stage 2a; 20 patients (36.4 %) in stage 2b and 8 patients (14.5 %) in stage 3 (III). From our patient pool, the period between the establishing of the diagnosis of MG and the period performance of thymectomy were as follows: 1) within 1 year of diagnosis (33 patients, 20 %); and 3) within 3 years or more of diagnosis (11 patients, 20 %).

Histologic typing of thymic tissues removed intraoperatively are as follows: a) thymic hyperplasia (28 patients, 50.9 %); B) thymic persistency (15 patients, 27.2 %); c) thymic involution (5 patients, 9.2 %); and d) thymomas(7 patients, 12.7 %). In the literature it is estimated that 8 to 15 % of patients with MG have thymomas, and approximately 30 % of thymomas are accompanied by MG (18-20). From our own patients had thymomas (which seems to confirm this estimate).

As show in figure 5, our postoperative results are as follows: a) remission (21 patients, 38.5 %);

b) asymptomatic (19 patients, 34.5 %);
c) improved (6 patients, 10.9 %);
d) unchanged (4 patients, 7. 3 %);
e) worse (2 patients, 3.5 %); and
f) died (3 patients, 5.5 %) (28).

The total percentage of patients (who either went into remission plus were asymptomatic plus improved) = 83.9 %, compared with 16.4 %, whose conditions were either unchanged, or got worse or died.

It is reported that about 30 % of patients are cured as a

result of properly indicated total thymectomy performed at the right time and that the condition of a further 30- 35 % improves considerably (21). The 38.5 % cure rate of our patients, after operation, approximates this literature value. This obviously could be attributed to a variety of factors. (considerable decrease in the proportion of anti-ACHR antibodies could serve as a marker (22). Out of all these factors, with the present technology and the state of our current knowledge, the literature considers the time elapsed between the onset of the disease and the operation to be the most important. It is regarded as optimal if the operation takes place within two years of the onset of the disease (23).

We believe this could be the single most important factor explaining the higher remission rate of our patients since 60 % of them were operated within 1 year of the diagnosis of the disease.

Conclusion

In concluding then, based on our experience, we suggest that operations (via the medial sternotomy approach) be done within 1 year of the diagnosis of the disease. The question about what to do with those whose condition are either unchanged or get worse despite surgery is a rather vexing one but which nevertheless must be addressed in time and appropriately solved.

The search for newer therapeutic modalities or more effective adjuvant therapy for this subgroup must continue unabated. For the time being, however, total medianostomy will continue to benefit a bulk majority of, but unfortunately not all, myasthenic patients.

APPENDIX I: LETTER OF PERMISSION

Jann Ferenc Korhaz-Rendelointezete
Organikus Ideggyogyaszati Osztaly
Ost vez foorvos Prof dr Szobor Albert
1204 Budapest xx. Koves u. 2-4.

Lawrence A. Zumo
Debrecen 12, Pf. 52. H-4012

Dear colleague,

After your request, I give you the permission to use the pg. 32. of my monograph, Myasthenia Gravis. Furthermore, you can use of course, the data of this monograph according to the customary citations in the literature.

I wish you a lucky diploma work, and later on much and good results and enjoys in your profession.

Budapest, 1, 12, 1992.

With personal greetings

Prof. A. Szobor
Osztalyvezeto , Ideggyogyaszati Intezete

REFERENCES

1. Hokkanen, E. (1969): Epidemiology of myasthenia gravis in Finland, J. Neurol. Sci. 9, 463-478

2. Besznyak, I., Szende B. Lapis K. (1985): A mediastinum tumorai es pseudotumorai, Akademia Kiado, P. 79

3. Szobor A.(1990): Myasthenia Gravis, Akademia Kiado, vii

4. Passmore, R., Robson, J.S (eds.) =1974): A companion to medical Studies, vol 2, p. 34, 31

5. Szobor A.(1990): Myasthenia Gravis, Akademia kiado,

6. Szobor A, (1990) Myasthenia Gravis, Akademia Kiado, p. 34 (culled with pemission)

7. Clarke, D.B. (1990) in: Givel, Jean-Claude (ED), Surgery of the thymus, springer-verleg, pp. 257-258

8. Drachman, D.B (1987); Present and future

treatment of myasthenia gravis, N. Engl. J. med. 316, 743-745

9. Szobor A. (1990): Myasthenia Gravis, academia Kiiado, p. 11

10. Merlini, M.and Clarde, D.B. (1990): Surgical approaches in: Givel, Jean-Claude (ed), Surgery of the thymus, Springer-verlag, pp. 249-252

11. Szobor A., Samu Zs. (1984): Ideggyogy. Szle, 37, 241-249

12. Szobor A. (1990): Myasthenia Gravis, Akademia Kiado, p. 17

13. Smithers, D.W. (1959): J. Fac. Radiol., 10, 3-16

14. Mumenthaler, M. (1990): Neurology, Thieme Publishers, p. 488-489

15. Mumenthaler, M. (1990): Neurology, Thieme Publishers, p. 488-489

16. Mumenthaler, M. (1990): Neurology, Thieme Publishers, p. 492

17. Yamamoto, T. et al. (1986): Anti-filamin and vinculin antibodies in sera from patients with myasthenia gravis and polymyositis, Proc. Jap. Acad. 62(B), 113-116

18. Rosai, I., Levine, G.D. (1976): Tumors of the thymus, atlas of tumor pathology, 2nd series Washington DC, Armed forces institution of pathology

19. Gray, G.F., Gutowski, W.T. (1979): Thymoma: A clinico-pathologic study of 54 cases. Am. J. Surg. Pathol. 3, 235-249

20. Wilkins, E.W. Jr., castleman, B. (1979): Thymoma: A continuing survey at the Massachusetts General Hospiital. Ann. Thorac. Surg., 28, 252-256

21. Szobor A. (1990): Myasthenia Gravis, Akademia Kiado, p. 160

22. Vincent et al. (1983): Acetylcholine Receptors antibodies and clinical response to thymectomy in myasthenia gravis, Neurology, 33, 1276-1282

23. Szobor A. (1990): Myasthenia gravis, Akademia Kiado, p. 160

24. Jaretzki, A ., A. Penn, A.S. et al. (1988): "Maximal thymectomy for myasthenia gravis." J. Thorac. Cardiovasc. Surg., 95, 747-757

25. Dalto, S.K., Schwartz, R.S. (1974): Infectious myasthenia, N. Engl. J. Med. 291, 1304-1305

26. Maggi, G., Cassadio, C., Cavallo, A. (1991): Thymoma: Results of 241 operated cases, Ann. Thorac. Surg., 51, 152-156

27. Priskkamen, R., Tiilikaimen, A., Hokkamen, E. (1972): Histocompatibility (HL-A) antigens associated with myasthenia gravis, Ann. Res, 4, 304-306

28. Kollar S., Hololay p., Peterffy A. (1991): Thymectomy in the treatment of myasthenia gravis. 5th Annual Meeting of the European Association for Cardio-thoracic Surgery (London) Abstracts, p. 204

CRITIQUE OF M.D. THESIS

Lawrence Amos Zumo

THYMECTOMY IN THE TREATMENT OF MYASTHENIA GRAVIS (A CASE STUDY OF 55 PATIENTS)

The construction of thesis fills the requirements. It consists of 33 pages including references (28 titles) and 5 figures. The position of figures is very special. Generally they are put into the text or they (and the legends) are attached to the end of the papers. The first half of the work is a literary review, the second half is the evaluation of clinical data of 55 patients operated in the Cardiothoracic surgery.

My English is not so good as to criticize the style and grammatic of the paper but it seems correct.

In the introduction the author summarises the epidemiology of the myasthenia gravis. To mention the distribution of patients according to gender and the short summary of results are unnecessary here, and these data are repeated later. A clear formulation of the aim of the study would have been advisable in this section.

The historical survey is a very skilful and concise summary of the most important stages in the recognizing of basic features of the disease. The section of surgical technique would have had a more correct place in the second half of the paper, after patients and methods. i.e. the technique mentioned here was used for the operation of patients referred to in this paper. The author did not take part personally in the operation, consequently I offer to use passive structures instead of plural I.

Sometimes the author speaks about resection. It is false, the aim of surgical intervention is the exstirpation of the gland.

The widely used cholinesterase inhibitor, the Mytelase (ambenonium), was not mentioned. It is important to know, that the demand of cholinesterase inhibitors varies in a very wide range inter-and intraindividually, too (e.g. from 2 tabs to 15 tabs Mestinon daily).

I would like to read a little more about the details of thymectomy: the history of the operations the general indication and contraindication of thymectomy in MG, the basic rules of intervention. It would be better to discuss these questions in a separate section. In the elderly patients the thymectomy is not absolutely contraindicated. The literature contains only a small number of reports concerning elderly patients with MG who underwent thymectomy. According to the few large series the results of thymectomy proved to be favorable even over 60 years of age.

Other autoimmune diseases are found not only in the relatives of MG patents, but in the patients itself. It was the first finding which have called the attention to the autoimmune pathomechanism. (pages 18.)

Although theoretically the direct blocking effect of the ACHR-s may cause transmission block, it does not play any significant role in performing the clinical features. The amount of antibodies directed against the ACHR binding site is only 1 % of circulating antibodies, and it can not induce the proved increased degradation and decreased resynthesis of receptors. (pages 18, 19)

The evaluation of clinical data of patients operated in the Cardiothoracic surgery is a valuable part of the thesis. The figures are clear and well constructed. The description of results are appropriate. But there are few points missing: 1. What are the surgical complications? 2. Were there postoperative myasthenic crises? 3. The details of case history of patients, who died after surgical intervention would be very interesting and instructive. 4. what are the anesthesiological requirements. Of patients with MG? 5. It would be important to discuss in detail the reasons of lack of improvement after thymectomy.

In summary; the paper is a good survey on the role of thynectomy in the treatment of patients with MG. it has a valuable evaluation of the experiences with thymectomy in patients operated in the Dept. of Cardiohtoracic surgery. My appreciation: very good (5).

Dr. Dioszeghy Peter
Debrecen, Hungary July 5 1993

Peter Dioszeghy, M.D., C. Sc.
Dept. of Neurology & Psychiatry
University Medical School
Debrecen, Hungary

Sir,

Your critique of my M.D. thesis has just been forwarded to me by the university's a thesis evaluation committee. Your comments and questions have been well received.

First and foremost, I wish to extend to you my heartfelt thanks and appreciation for the preoperative check-ups of most of our patients, your advices and your critique. To your queries, the following points and clarifications are thus in order:

a) Your assertion that " I did not personally take part in the operations " is a correction on a technicality. I personally participated as 2nd surgical assistant in some of the operations (i.e. 1990 to the present) albeit intermittently due to time conflicts and other pressing engagements. Since the "MG surgical team" has since 1981 (the beginning of our current sampling) employed basically the same surgical approach, I feel my post – 1989 surgical experiences justify my original statements. (for purely arbitrary reasons, we chose to do a 10 yr-, instead of a 13 yr-, sampling. Had the latter been undertaken, this would have clarified matters).

b) Due to severe university page restrictions and other thesis regulations, it was quite difficult to decide how detailed this work should be. This is why interesting topics like; history of the operations, newer surgical innovations, general indications, in MG, basic rules of intervention, detailed correlations between thytmic pathology and results of operations in those over age 60, etc were excluded. Unfortunately, this may not have done justice to the subject.

c) The question concerning the increased degradation and decreased resynthesis of receptors, to date, remains a challenging yet interesting one-considering the small fraction of circulating anti-ACHR Ab's purported to be involved. In concurrence with your opinion, Zimmermann and Eblen (Clin. Investing. (1993), 71; 445-451) in demonstrating that the repertoires of autoantibodies against homologous eye muscle in ocular and generalized myasthenia gravis differ and thus lending support to the accumulating evidence in favor of an etiologic heterogeneity of MG, although via circumstantial evidence, point to the direction that other configurations and consorts of antibodies possibly additional target sites, etc. may be involved. They cautioned that these speculations must await perhaps the refinement of our monocausal concept of the pathogenesis of MG and also that of the development of appropriately refined or newer technologies to enable these questions to be definitively put to rest. Conceivably, a spin-off from any number of these advances could help us explain eg. Why there was a lack of improvement post-thymectomy in some of our patients.

d) Question 1: common surgical complications:

a) blood loss when large mediastinal vessels are injured.
b) Air embolism due to subatmospheric. Pressure in lumens of mediastinal vessels.
c) Unobserved pneumothorax (eg. when small rupture in

pleura is not noticed and left unsutured).

d) Injury to the phrenic nerve during the preparation of the vessels.

e) Wound infection in the early postoperative period and mediastinitis. In our patient series, we had only 1 surgical complication, i.e. mediastinitis.

e) Question 2: Postop. Myasthenia crises: in our series,
there were 3 such crises. Adequate medical treatment and prolonged respiratory treatment controlled these episodes. In one of these 3 cases, we had to use plasmapheresis in the early postop period.

f) Question 3: Cause of death in postop patients. 3 patients died postoperatively. Of these, 2 died from causes unrelated to the surgery. The third patient's cause of death could be related to the surgery.

He was a 56 years-old male who had prolonged immunosuppressive therapy (eg. Corticosteroids) before surgery. This might have contributed to his mediastinitis. Despite use of local/general medical treatment (eg. Antibiotics), he died 6 months later from septico-toxic

causes.

g) Question 4: Anesthesiologic requirements for MG patients.

a) General aim: Locoregional anesthesia should be preferred whenever allowed by the patient's clinical condition and the type of operation, as the lack of intubation and the postoperative analgesia can be advantageous.

b) A common practice and as described in the literature, is to administer 25 mg hydrocortisone over the next 24 hrs (in addition to the usual dosage) if the patient has received corticosteroids for more than 1 month during the preceding 6-12 months.

c) An inhalation anesthesia induction is recommended for those with large thymic masses. Alternatively, awake intubation under topical anesthesia, with fiber optic bronchoscopy is a safely used technique. In extreme cases, extracorporeal circulation can be established prior to anesthesia (due to the tracheobronchial compressiuon, superior vena caval obstruction or even cardiac tamponade that may be caused by this pathology).

d) The condition of the patient should be optimal prior to surgery, particularly with regards to the control of myasthenia and freedom from respiratory tract infections. Hypokalemia is to be particulary screened and avoided as this will exacerbate myoneural weakness.

In severe myasthenia, plasma exchange is an important part of the preoperative preparation.

f) Premedications: opiate analgetics, benzodiazepine sedatives, and anticholinergic drugs.

g) Neuromuscular blocking drugs are best avoided altogether. (although the short acting, non-depolarizing blockingagent,atracurium,has been reported in the literature to be without complications when used in MG patients for non-thymic surgery).

h) Deep anesthesia is rapidly achieved with halothane. Medical therapy is reintroduced before extubation only if it is imperative for respiratory function. The straightforward treatment for respiratory decompensation includes intermittent positive pressure breathing (IPPB), bronchial aspiration and appropriate antibiotics.

i) Question 5: Speculations about the lack of improvement post-thymectomy in some of our patients. As important, interesting (and vital to construction newer therapeutic strategies) as this question may be, we feel that the search for its immediate answer (s) does not fall under the domain of our current surgical investigations. To the best of our knowledge, current research efforts in neuromuscular disorders and autoimmunity are being intensely devoted to addressing this very question more appropriately and perhaps a fruit of that search could provide more definitive answers.

Once again, thanking you for your critique, advices, queries and pre-op checkups of our patients and finally wishing you the very best of good fortune.

I remain,

Yours sincerely,

Lawrence A. Zumo
Trainee Intern

Cc: Thesis evaluation committee
Dr. Sandor kollar, Supervisor
File

Critical remarks

On the thesis of Lawrence A. Zumo

Lawrence A. Zumo submitted his thesis under the title of thymectomy in the treatment of Myasthenia Gravis. It is 27 pages including 5 pages of diagrams. In his references he refers to 28 literature data, which are not given in alphabetical order but in the order of references. Mention must be made that he refers only to 20 and not 28 different sources.

In the introduction he writes about the frequency of the disease and its distribution between sexes. In the third paragraph of page 3 the author could have mentioned the number and not only the percentage of the patients operated on within one year of diagnosis.

In the historical perspective the author describes the recognition and the first attempts at treatment of MG. in the second paragraph of p. 4 he mentions the names of Dr. Mary Walker, Dale and Feldberg, who published their initial research and treatment results. Where did the author take these data from? There is no reference to these names in the References and no citation of any other author.

In the third paragraph of p. 5 he mentions Erb's name and on p. 6. the names of Goldflam, Jolly and Walker as researchers describing the symptoms of MG and in his references we find Szobor's name. The appropriate form should be to cite Szobor in these cases too.

In chapter surgical technique the author describes the different approaches of thymectomy. It is clear and easily understandable.

In therapeutic Modalities the author describes the principles and possibilities of drug therapy of MG. This is one of the best and most useful parts of the thesis.

Chapter Pathology is accurate and correct. Though I miss histological diagrams and photos very much. They would have made this chapter much more valuable and interesting.

The material of a 10 year period in chapter results and discussion is good and correct. Mortality rate (5.5 %) is higher than in literature in general but it can be due to the lower number or patients and operations in later stages.

The author's choice of this topic is current. Apart from a fews lighter mistakes I suggest accepting and grading it excellent.

Dr. Nabradi Zoltan, adjunktus
Dote II. Sebeszet

Debrecen, Aug. 11, 1993

Nabradi Zoltan, MD
University medical school

Debrecen, Hungary

Sir:

Your critical remarks are finally at hand. Thanks for your analysis, suggestions and comments.

Those patients operated within 1 year were 33 in all (i.e. 60 % of total). Thanks for your correction.

More appropriately, the observation of Dr. Mary Walker, the work of Dale and Frldberg were referenced in: (passmore. R. Robson, J.S. (eds.) , (1974), A COMPANION TO MEDICAL STUDIES. Volume 3, part 2, p. 34. 31).

The works of Erb, Goldflam, Jolly and Walker were taken from Szobor's monograph, MYASTHENIA GRAVIS, (see refs. 3.5).

Additional references as cited by Szobor are:

1) Erb, W (1878-1879), Zur kasuistik der bulbaren Lehmungen. Uber einen neuen wahrscheinlich bulbaren symptomenkomplex. Arch. Psychiat. Nervenkrenkheit, 9,336-350.
2) Goldflam. S (1893), Uber einen scheinbar heilbaren bulbar-paralytischen symptomenkkomplex mit beteiligung der

Extremitaten. Dtsch. Z. Nervenheilk, 4.312-352.

3) Jolly. F(1895) Uber myasthenia gravis pseudoparalytica. Berl. Klin. Wochenschr, 32, 1-7
4) Walker, MB (1934), Treatment of myasthenia gravis with physostigmine, Lancet, I, 1200-1201.
5) Walker, MB (1938), Myasthenia Gravis. Case in which fatigue of the forearm muscles could induce paralysis of extraocular muscles. Proc. R. Soc. Med. 31, 722-723.

I take your comment about dearth of histologic diagrams and photos fully without further ado. Due to general unavailability of the above mentioned and difficulty in obtaining photo credits ere these culled from textbooks, I decided to leave them out of the thesis. In the future, however, particular attention will be paid to this omission.

Once again, thanking you for your critical remarks, suggestions and analysis.

I remain,

Sincerely yours,

Lawrence A. Zumo
Trainee intern

Cc; Thesis Eval. Comm..
Dr. Sandor Kollar
File

MINUTES OF THESIS DEFENCE

Student's name:Lawrence Amos Zumo
Title of thesis:THYMECTOMY IN THE TREATMENT OF MYASTHENIA GRAVIS"

Name of supervisor: Sandor Kollar M.D.
2nd Department of surgery

1 referee: Pete Dioszeghy M.D. C. Sc.

2 referee: Zoltan Nabradi M.D.

Short assessment:

Lawrence Amos Zumo gave very correct answers to the questions being put by the referee after his summing up his these" thymectomy iin the treatment of myasthenis gravis"

The committee found the thesis excellent both by its form and contents, and accepted it as an excellent (five).

Final mark: excellent (5)

Debrecen, 1993, 08. 13...............
Chairman of committee

PART V:

POST MEDICAL SCHOOL PERIOD

At the end of medical studies, I migrated to the USA to join my wife and new born son who was then just 5 months ; and to validate my studies, prove my worth on the world scale. After US licensing exams and a year's stint as a research program coordinator at the Johns Hopkins University School of Public Health in Baltimore, Maryland, I landed a position as a medicine intern in New York. After a year, I went on to do my neurology residency and fellowship in autonomic nervous system disorders.

All along I keep up my scientific research interest. This time taking a clinical case and doing the basic science research on it. Here are a examples of that effort.

THYROTOXIC HYPOKALEMIC PERIODIC PARALYSIS

IN A HISPANIC MALE

Lawrence A. Zumo, MD (1), Christian Terzian, MD (2), Timothy Brannan, MD (3)

(1) JFK Neuroscience Institute, JFK Medical Center, Edison, New Jersey
(2) Dept of Medicine, Jersey City Medical Center, Jersey City, New Jersey
(3) Director of Neurology, Jersey City Medical Center, Jersey City, New Jersey

THYROTOXIC HYPOKALEMIC PERIODIC PARALYSIS

IN A HISPANIC MALE

ABSTRACT:

We report a case of a Hispanic male presenting with acute onset of bilateral lower extremity weakness, without any antecedent viral or bacterial illness, dietary changes as well as no infiltrative orbitopathy, diffuse goiter, infiltrative dermopathy nor family history of periodic paralysis, who was later found to have Graves' disease. This demonstrates a rare case of periodic paralysis as initial presentation of hyperthyroidism. THPP is

common in Orientals and uncommon in Caucasians and African Americans, but does occur in individuals of Hispanic origin. Copyright. (J. Nat. Med. Assoc. 2002;94:383-386)

INTRODUCTION:

The several forms of periodic paralyses are either primary or secondary forms and the pathologic

features are characterized by episodic attacks of flaccid paralysis. The primary or familial paralysis are

divided into hypokalemic, normokalemic and hyperkalemic varieties; and the secondary periodic paralysis

are associated with thyrotoxicosis, primary aldosteronism, abnormalities of potassium in diabetic acidosis and renal tubular acidosis (18).

<u>Hypokalemic Periodic Paralysis</u>: This is the most common form, which is transmitted by an autosomal

recessive gene but is three times more frequent in males than in females. Attacks of paralysis are

characteristically nocturnal and usually occur after a

period of vigorous exercise. Symptoms usually appear

first in the second decade and tend to be less severe with age. Our case is that of a Hispanic male,

who, although of the expected ethnic group (i.e. immigrated from El Salvador), lacked the clinical

symptomatology except for subtle fine hand tremors.

Given the clinical setting, the nature of presentation, the physical examination and the

paraclinical studies, diagnostic entities that must be ruled out include: hypokalemic or

normokalemic periodic paralysis, Guillain-Barre syndrome, myasthenia graves, Eaton-Lambert

syndrome, multiple sclerosis, transient ischemic attacks, metabolic defects of muscle (impaired)

carbohydrate or fatty acid utilization), chronic renal failure and surreptitious diuretics abuse.

Although associated with low levels of serum potassium, weakness may begin at potassium levels higher than those that would include paralysis in a normal individual.

CASE REPORT:

A 41 year old Hispanic male (who immigrated from El Salvador) with no significant medical, social

or family history was brought to the emergency room of the Jersey City Medical Center, New Jersey

complaining of acute onset of first episode of bilateral leg weakness with inability to stand or ambulate

when he awoke from sleep in the early hours of the morning. He reported no associated symptoms of

focal numbness, dysarthria, pain, diplopia, impairment of consciousness nor sphincter

incontinence. He denied any recent dietary or medication changes, no previous strenuous

exercises, nor antecedent viral infections. Patient denied diarrhea, cough, fever, constipation, hoarseness of voice, lethargy, abdominal pain, palpitations, recent changes in weight or appetite

or alteration of mood.

Patient had no known drug allergy. He was only taking hydroxyzine (for urticaria on his legs).

Family history is not significant for muscle weakness.

Approximately 3 hours after the onset, his examination in the emergency room, showed

That he was in no acute cardiorespiratory distress and his vitals were: blood pressure of 163/81,

Heart rate of 95 beats per minute, respiratory rate of 12, temperature of 97.8 deg. F.

General examination: HEENT: atraumatic, normocephalic, no jaundice, no ear or nose

discharge. Neck: supple, no JVD, no lymphadenopathy, no thyromegaly, no thyroid bruits or

nodules. Chest: clear to auscultation. Cardiovascular

system: S1, S2 regular, 2/6 systolic murmur,

no gallop. Abdomen: soft, nontender, no organomegaly, positive bowel sounds, vitiligo on the

lower abdomen. Extemities: no clubbing, cyanosis or edema, and pulse palpable in all limbs.

Initial neurologic examination was significant only for bilateral lower extremities weakness

(power of 3/5 proximal and distal) with intact reflexes, negative Babinski bilaterally and fine tremors

of the outstretched hands.

Repeat neurologic examination three hours after the initial exam revealed: Neck supple,

No bruits, no meningeal signs. He was alert, oriented to person, place and time with intact

higher cortical function. Memory: 3/3 x 5 min. Speech: clear. Language: fluent. Visual field fullto finger counting. PERRLA (4-2mm). No afferent pupillary defect. VA: 20/20 bilat without correction. Fundus: no papilledema, no atrophy. EOM: intact. No nystagmus. No ptosis. Corneals

intact bilaterally. No facial asymmetry. Hearing intact. Gag intact. Tongue is midline. No

fasciculations. Motor: normal bulk and tone, with no evidence of atrophy, fasciculation, or

spasticity. Strength: 5/5 x 4 extremities. Reflexes: 2+ symmetric. Toes: down going bilaterally.

Normal anal sphincter tone. Sensory: intact light touch, temperature, pinprick, vibration and joint

position. Coordination: intact finger to nose, heel to shin. On the outstretched hands, patient had

bilateral fine tremors. Gait and stance normal.

The initial blood profile showed Na=131, **K=2.3mEq/l**, Cl=102, HCO3=22, glucose=

110, BUN=20, Cr=0.4. Total CPK = 103. WBC=8.5, Hgb=13.6, Hct=40.5, Platelets=200,

RDW=12.4, MCV=86.6.

The electrocardiogram showed normal sinus rhythm at 80 beats per minute. There was no

QT prolongation, no U waves nor ST segment depression.

Management plan included potassium replacement and IV hydration. Patient was supplemented

With potassium 40 mEQ orally and 40 meq IV over 4 hours of KCL. After four hours, repeat labs

Showed **K=4.2**. Patient, however, regained his full muscle power before the completion of the potassium

load.

Thyroid function tests revealed biochemical hyperthyroidism, with the following laboratory

values: TSH=0.01, T3=301 (59-174), T4=37.5 (4.5-12). Other laboratory values: Serum aldosterone =

7.0 ng\dl (1-16), Urine osm=694 (500-850),

Na=165, K=48.1, Cl=147, 24 Hr Urine collection: volume

5500ml, Na=649 (110-250), K=94, Cl=616 (110-250).

A thyroid scan was performed. Patient was given 270 uCi I-123 orally and a thyroid scan

Performed with anterior, right and left lateral oblique views 24 hours later. The radioactive iodine

uptake was 50% at 2 hours and 83% at 24 hours. The thyroid glands were of normal size and

configuration and there were no cold nodules. Findings were most consistent with Graves'

disease. Patient was started on Inderal 20 mg BID initially but later received one standard dose of

radioactive iodine I-131 and thyroid function values eventually returned to normal. He remains

euthyroid with no further episodes of paralysis reported. Attempts to do genetic analysis via a research/

genetics laboratory was unsuccessful due to several technical and logistic factors beyond our control.

DISCUSSION:

THPP is a rare complication of hyperthyroidism in the United States. It is most often

reported in Asians and rarely in Hispanics and less so in African Americans. Males to female ratio

is more than 1:12 and Graves' disease is more commonly reported in females. THPP generally

occur in the third to fourth decade of life. It can occur in hyperthyroidism, even secondary to

exogenous thyroid hormone supplementation-occurring several months to years after the onset of

hyperthyroidism or it may be the first symptomatic presentation of Graves' disease (8,9,11,15).

The paralysis is similar to Familial Periodic Paralysis-

FPP- but is not hereditary. Hypokalemic periodic paralysis, an acute intermittent muscle weakness (1), can occur in patients with thyrotoxicosis, most commonly in Oriental or Hispanic males, where up to 10 percent of thyrotoxic patients may have periodic paralysis. The pathogenesis of thyrotoxic periodic paralysis is uncertain but there is evidence of decrease in the activity of the calcium pump (2,3,4).

Associated with the diverse causes of extracellular potassium deficits, this entity may often go unrecognized or misdiagnosed, while periods of normal functioning and normokalemia betweeen attacks serve to compound the diagnostic challenge (5,6). One, however, must retain a high index of clinical suspicion and exclude other neurologic/medical diagnostic entities that my have similar presentation but have varying severity of clinical outcomes and differing management approaches which must be decide upon as quickly as possible. Hyperthyroidism is not always clinically apparent and may be subtle or even subclinical (7).

Acetazolamide which is used for prophylaxis in FPP worsens THPP (11, 12). Both entities occur

as sudden onset of weakness lasting 3-24 hours or more, with some patients experiencing

prodromal symptoms such as muscle cramps. The legs are generally more involved that the arms

and proximal muscles are involved more often than the distal musculature. Reflexes may be

diminished or absent (and this may complicate the differential of Guillain-Barre syndrome).

Bulbar, ocular and respiratory muscles are not usually involved, but if so, the attack may be fatal.

Sensory and mental functions are spared.

The mechanism of THPP is not well understood, but it is postulated to be secondary to a shift

of potassium to the intracelluar space secondary to increased activity of the Na\K pump and beta

adrenergic hypersensitivity. Potassium causes depolarization of the resting membrane until the

sarcolemma become electrically inexcitable and thus paralysis occurs.

In support of this mechanism is the experimental evidence that epinephrine can induce paralysis in

Patients with THPP and the nonselective beta-blockers have a protective effect. Additionally,

clinically it has been observed that increasing carbohydrates in the diet can can increase insulin

secretion which stimulate the Na|K pump and exacerbate an episode in a thyrotoxic patient (14).

There is no relationship, however, between the severity of hyperthyroidism and the frequency

and severity of paralysis.

There is no loss of potassium and, as was evident in our case, the paralysis can be corrected

even before the completion of potassium supplementation because potassium flows back outside

the cells. Treatment should first be with supplementing potassium to reverse the paralysis and to

protect the heart from the effects of hypokalemia while carefully monitoring the serum potassium

to prevent hyperkalemia from the outflux of potassium from the cells.

Then a nonselective beta-blocker such as propranolol should be started until the hyperthyroidism

is corrected and once thyrotoxicosis is under control, the paralytic episodes will cease (13). There

is a high recurrence rate of hyperthyroidism after long term antithyroid therapy, hence early

radioiodine ablation of the thyroid was instituted in our patient (6).

In our case, potassium was supplemented immediately to prevent life threatening arrhythmias before hyperthyroidism was confirmed by laboratory findings, as the patient had only minimal symptoms and signs of increased thyroid functions. In one retrospective study, supplementation of potassium resulted in rebound hyperkalemia in more than 40% of the cases (19).

In a recent study, oral administration of propranolol at a dose of 3 mg/kg increased serum potassium and phosphate concentrations in two hours without rebound hyperkalemia and hyperphosphatemia (20).

REFERENCES:

1. Gavin GM: Sudden onset of weakness in a 24 year old. Postgrad. Med 1990; 88: 109-110.

2. Isselbacher, Braunwald, et al: Harrison's Principles of Internal Medicine; Periodic Paralysis, Ch. 387, p.2396, 1994 13th Edition.

3. Ober, KP: Thyrotoxic periodic paralysisin the United States: Report of 7 cases and review of the literature. Medicine 71: 169,1992.

4.Engel AG: Periodic paralysis, in Myology, 2d ed, AG Engel, Franzini-Armstrong C (eds.), New York, McGraw Hill, vol.2, 1994.

5. Jackson CF, Barohn RJ: Improvement of the exercise test after therapy in thyrotoxic periodic paralysis. Muscle Nerve 1992; 15: 1069-1071.

6.HochbergDA,VassoloM,PaniaguaD:Thyrotoxic periodic paralysis in a black man. Southern Medical Journal. 89(7): 735-7,1996 Jul.

7.Ober, KP: ibid.

8.See Reference 6

9.Ober KP: Thyrotoxic Periodic Paralysis in the United States. Medicine. 1992, 7:109-120

10. Lawrence GD, Chwa E, Balagtas R: Thyrotoxic periodic paralysis. Md Med J 1990; 39: 583-587

11. Capobianco DJ. Hyperthyroidism and periodic paralysis, J Fla Med Assoc 1990; 77:884-888

12.Shulkin D, Olson BR, Levy GS: Thyrotoxic periodic paralysis in a Latin American taking Acetazolamide. Am J Med Sci 1989; 297:337-338

13.Chan A, Shinde R, Chow CC, et al: In vivo and in vitro sodium pump activity in subjects with thyrotoxic periodic paralysis. BMJ 1991 ;303:1096=1099

14.Conway MJ, Seibel Ja , Eaton P: Thyrotoxicosis and periodic paralysis, improvement with betablockers. Ann Intern Med 1974;81:332-336

15.McFadzen AJS, Yeung R: Periodic paralysis complicating thyrotoxicosis in Chinese. Br Med J1967;1:451

16.Saeian K, Heckerling PS: Thyrotoxic periodic paralysis in a Hispanic man. Arch Intern Med 1988,148:708

17.Stedwell RE, Allen KM, Binder LS: Hypokalemic paralysis: A review of etiologies, pathophysiology, presentation and therapy, Amer J Emerg Med 1992; 10:143

18.Riggs JE: The periodic paralyses. Neurol. Clin. 6(3): 485-98. Aug. 1988.

19. Manoukian MA: Clinical and metabolic features of thyrotoxic periodic paralysis in 24 episodes. Arch
Intern Med 1999 ; 159(6): 601-6. March 1999.

20.Lin SH: Propranolol rapidly reverses paralysis, hypokalemia and hypophosphatemia in thyrotoxic
periodic paralysis. Am J Kidney Dis: 37(3): 620-3. March 2001.

With the emergence of new techniques in molecular genetics and the immense effect it was poised to have on different aspects of clinical medicine, especially neurology and neuroscience, I had to the great fortune to work in the laboratory of Dr. R. Grewal, neuromuscular specialist, during my last semester of residency training. The result of that collaboration culminated in the publication of a clinical neuroscience article, which is excerpted below.

Castleman's disease-associated neuropathy: no evidence of human herpesvirus type 8 infection.

Zumo L, Grewal RP.

Laboratory of Neurogenetics, New Jersey Neuroscience Institute at JFK Hospital, 65 James St, Edison, NJ 08818, USA.

Castleman's disease (CD) is a rare lymphoproliferative disorder that may be associated with a neuropathy. In a recent report, the presence of human herpesvirus type 8 (HHV-8) DNA sequences were detected in an HIV-negative patient with polyneuropathy, organomegaly, endocrinopathy, M-protein and skin changes (POEMS) associated with the multicentric hyaline vascular variant of CD. It was proposed that the presence of these sequences may have a role in the pathophysiology of the neuropathy.We describe an HIV-negative woman with the multicentric plasma cell form of CD who presented with a disabling neuropathy. In addition to a severe demyelinating polyneuropathy, she had some of the other features of POEMS including an IgA lambda gammopathy and lymphadenopathy. We were unable to detect the presence of either HHV-8 DNA or proteins in this patient. The significance of our results and the relationship between CD, POEMS and neuropathy are discussed.(J Neurol Sci. 2002 Mar 15;195(1):47-50).

PART VI:

POST TRAINING PERIOD

A critical look at Liberia's direction and general suggestions after the immediate cessation of prolonged fracticidal civil war. Below is the full text of my address to a gathering of Liberians at the North Carolina A& T University in October 2004.

IMPERATIVES FOR COLLECTIVE SECURITY IN THE NEW LIBERIA

Liberia is at a critical juncture. We have just come thru a catastrophic convulsive seizure and profound post-event confusion still persists. Where we go from here will depend on the rational choices we make now and how we collectively, openly and sincerely carry them out. Other nations have been here before. Some succeeded and others failed miserably. We must agree to disagree ,if need be, but we should keep our eyes and minds on the best interests of our nation or our collective demise is certainly assured. The task at hand require solid inner fortitude, extreme discipline, foresight and a level of patriotism never before experienced in our nation. Recent painful examples should forever remind us that despite all the rhetoric and apparent paradox nobody else from outside will help us more than we can help ourselves. They all have their own agenda. Do the math! All of us cannot attack the same problem the same way. Or else our efforts and results will only be limited But we must each do or say something, no matter how unpopular it may be. What may be popular is not always right and what is right is not always popular. The system that currently exists in Liberia does not work. We need new bold thinking and action. It is in this vein, using our utmost imaginations and analytical skills, I wish to offer you some practical suggestions concerning the task at hand for our nation.

1)I will propose that we seriously consider, after due deliberations, establishing an **Economic & Security Crimes Act**. Because the importance of this Act, it should be enacted by the Legislature and not passed by Presidential Decree as it should survive each succeeding president because this of perpetual national interest. Under this act, we will deals with eg. issues of businessmen and their deals/ negotiations that are directly or indirectly detrimental to our national security interest and long term economic goals. The issues of bribes, tax evasions, misappropriation/embezzlement of funds, shadowy loans, etc will be tackled under this act. It is recommended that enforcement of this Act should fall under the jurisdiction of a panel of Superior Court Judges. Sentencing guidelines will have to be established to ensure punishments commensurate with the crimes are levied appropriately. Decisions concerning hearings in private or public giving national security concerns are to be debated. Open or anonymous tips and suggestions from all citizens about all activities at levels of government and individual activities. Whistleblowers protection to be guaranteed. Legal rights and representation of people so accused are to be safeguarded.

2)The Establishment of an **External Economic Intelligence Service (EIS for short)** whose function will be to gather, process, appropriately disseminate useful economic/ manufacturing mechanisms and processes from places we have visited abroad that we can readily adapt to our setting for rapid economic integration as part of our long term strategic goal. This Service should be a vibrant part of our present Ministry of Economic and Planning Affairs. This Service is to be different from the present stifling, counterproductive

politicized internal intelligence agency whose true values we are yet to realize as a nation. No need for endless ministerial delegations. Our embassies abroad, students abroad and patriotic citizens should make this reality.What are our embassies and missions abroad for? Private country club for a privileged few?

3)All previous, present and future **government loan agreements** by our government should be published in full for all to see so we all can review the repayment terms, the signatories to those loans, etc.

4)Competitive inter-county fairs annually to showcase agricultural products, livestock and local ingenuity as well as annual competitive national/inter-county high school science fairs with different prizes to vie for inorder to encourage the scientific education and manufacturing processes early. Selection of teams of talented Liberians who have won at the national level to participate in such events abroad for broader recognition. Hence, the restructuring of our national awards system (Star of African Redemption,etc) to go to Liberians at home and abroad who make meaningful contributions to our nation to be so recognized via a free, transparent and fair selection process. What can we learn from the English and South African award systems?

Liberia, via successive inept leaders, has not made us proud. No matter how you look at it, no formula for success exists in Liberia. Liberia has abandoned its children. How can you love a parent that has abandoned you? No national prioritization in the awarding of limited government scholarships exists. Talented students sent abroad for further studies are habitually abandoned in midair at the whim and wishes of the president and/or his ministers whilst ministerial girlfriends and their relatives are on fully paid government scholarships to study subjects like secretarial science, African studies, underwater basket weaving and art design. Many of these 'special students' go on to be superannuated students and guess who foots the bill? Simultaneously students who study in nontraditional countries, many very qualified, upon their return home are discriminated against to the point of nonexistence. When they leave Liberia in anger, then their patriotism is perpetually questioned. Yet some of us try to be patriotic and that is why we are here today.

Instead of sending numerous ministerial girlfriends

on multiple shopping sprees and escapades as well as countless presidential and ministerial health checkups abroad without any clearcut benefit to the population at large, we could reinvest most of the money spent this way in the EIS. Despite the abundance of our natural resources, yet we do not have a manufacturing base nor psychological willpower nor national pride to decrease our perpetual dependence on foreign goods. I am sure we can benefit much as a nation if we catalogue what we can learn from the Jamaican apprenticeship system, Hungarian farming and manufacturing efficiency methods, Brazilian local technological transformations, Ghanaian educational models, South African energy production techniques, German and Japanese engineering capabilities, Israeli wealth generation and entrepreneurial skills, Chinese/ Iranian and Ukranian offset printing and electronic reproduction techniques.

Our market women, despite their lack of formal education, have a lot to teach us. They have time tested skills, ingenuity and wisdom that we can draw upon as we embark upon the very tortuous road but many of us are too proud or even shy to be seen associating with them. Let me remind you that whilst others were involved in wanton corruption and mortgaging away our future, many market women sold their local produces and other goods and even sent some of their children abroad for further studies. They are experts in their own way. We need to turn to them for their input. It has even been suggested that we need to have a representative of the Liberian Market Women Association in every ministry for their input.

Our products, achievements, culture, etc should be displayed at our embassies and missions abroad during Independence Week without exception.

From many indicators that we read, we are told that we are behind by 75-100 years. If we want to close this within 15-25 years, we have to seriously look at means of wealth generation that white Americans, Europeans and Japanese have mastered so well. This quest goes hand in hand with a strong educational base. China, Brazil, Korea and Singapore are well on their way in this regard. Atta Turk of Turkey, David Ben-Gurion of Israel, Julius Nyerere of Tanzania, Jerry Rawlings of Ghana, David Suzuki of Japan, Nelson Mandela of South Africa, Hugo Chavez of Venezuela and Mao Tse Tung of China ,etc–all had vision for their countries and they delivered. Could we not do the same for our country?

We have most of these structures in place with out much structural adjustment. Instead we need more mental/intellectual adjustment. Look at our embassies abroad currently. In my opinion, they can do more for Liberia and Liberians. What are they for? Country club for the privileged few?

Things we need to look at critically and change within ourselves if we are to make any true strides towards our goals:

Mental Lethargy (Oblomov) : We have ample sugar cane supply but we still import sugar; we have cassava but still import flour, etc; we can make and export quality whisky and wine from sugar cane, rice, etc; yet we spend enormous money on importing whiskey; we have coconut, palm nut, etc; yet we die to import cooking oil. We have "jologbo" for worms, yet we yearn for worm medicine from abroad; we have *Raulwolfia* trees with abundant reserpine, yet are dying waiting for imported high blood pressure medicines, without considering investing in local natural products chemistry; we have abundant mango, guava, etc but we are dying to buy jam from abroad; oranges and other citrus fruits; yet we await juices from abroad; we have trees and wood but we want imported furniture; abundant fish but can't wait to buy canned sardines; abundant coco and coffee but we can't wait to get our morning ovaltine; etc, etc.

We have printing presses at home. Why could we not print nice, attractive labels to put on our containers/ glass jars for these products.

Go in any USA grocery store and see how many

labeled made in USA? Go in any Liberian grocery store and see how many made in Liberia?

The ministry of Commerce thru the World Trade Organization should be the one to deal with negotiations concerning the unfair trade practices and subsides by the developed countries once we are in the exporting mode. We can learn how the people of Tunisia and Mauritius are doing it.

You are prouder when you have something made in your country that many other people use and desire.

Co ngo-and-'Congolites'- better than Natives mentality: The debate rages on.

Consummate alcoholics mentality L: No amount of voluntary self policing has worked so far even if we are hit hard by calamity, punishment or tragedy; yet it seems that no learning has taken place.

Abdication of Parental Responsibilities: Parents are not setting good examples. We all can cite numerous examples off the top of our heads. For example, Liberian males having kids and basically leaving them to fend for themselves and their poor mothers always bear most of the burden of child upbringing as well as the blame for their eventual failures. We often hear: 'my parents did not do this for me, so be thankful." If they are your kids, you have to be responsible all the way. Not halfway.

Abdication of Financial Responsibilities: Paying bills and taxes to the teeth in America but then when we go to Liberia, we refuse to take responsibility for all these things; i.e. trying to have it both ways and think it is alright ; yet we wonder why we are where we still are.

Pretty Floyd Mentality: Dressing tip to top with nothing in our brains nor pocket and proudly showing off to others while our household members are starving and disintegrating. Always putting others down and ridiculing those who are working hard and trying to make a living the old fashion way by earning it.

Abuse of privilege: Abusing our offices as kings and not as public servants."Do you know who I am?" Jobs being awarded not to the best candidate but because of family, tribal, relationship connection or under the table deals. How can we then progress as a nation? No accountability.

Crabs &co and Back Stabbing Mentality : Pulling each other down like crabs because of fear they will gain more recognition that us. Where is our patriotism??? Where are we headed? Not helping another Liberian just for the sake of Liberian patriotism but always for our own ulterior motives.

Lack of respect for each other and each other's household/family because the person is poorer or uneducated.: Where are we going?

Money Mismanagement: Many of us are in the habit of spending money like there is no tomorrow. Save, Save, Save for your children's future, your retirement. Invest locally in eg. manufacturing ventures. Develop confidence in our banks with deposit insurance, so we can deposit money locally. So that money can be turned around as loans, venture capital, etc with returns/interests. American mostly deposit their monies locally, etc. You can make a lot of money but if you spend it lavishly without plans for the future, you will really never make it . We can learn a lot from the Liberian market women if we want to .

From Sonnewein to Beverly Hills Mentality:misplaced priorities. Living-above-our-means: when you do so, you tend to overextend, steal, lie, cheat and are more prone to corruption. The rest is history. Look at the Mexicans who share a single apartment instead of living large and in exclusive neighbourhoods so that they can save their money and open their own business or send money back home,etc.

I have a degree and everybody should adore me now mentality: What are you doing with that degree?

Don't play on my intelligence mentality: The saying goes: you can learn a lot from a dummy. Don't be offended when called stupid. Let the person tell you why. Maybe you can learn something from that person. Hello, common sense is not so common. Often, we hear: Don't tell me anything. We all know why we are in America. IS THAT REALLY TRUE?

Learning English Only Mentality: Treasure troves in other languages are being ignored to our peril due to our narrow-mindedness. Look at affordable Hungarian pocket sized junior and senior high school textbook summaries that most of us do not know about. How about the debate about a single language to supplement English in Liberia. The debate can be initiated. Swahili is quickly becoming the lingua franca of East and Southern Africa. There are intrinsic advantages there. What can we learn from them?

From America only mentality: We all know how far we have gotten with that mentality.

Somebody else can do it better than a Liberian mentality: We should know better by now from our travels abroad as they have shown us bluntly that theirs is better or more desirous no matter how inferior the quality. By this attitude, they usually help to improve the quality of their products. It is not bad to appreciate things from other countries but you should put things in perspective. Years ago, we used to laugh at Hongkong made but that is no more. What happened?

Finally, **the fatalistic; the you got to die from something Mentality:** God's will is not enough. It must be put in its proper prospective. It should not be an excuse for thinking. We have to begin to more seriously investigate things and to connect the dots more often. We must begin to ask questions-oral or written, no matter how dumb we feel they are.

In summary then, these are some foods for thought that I hope we will examine critically. **Along with economic prosperity comes collective social security, healthy population, longevity, etc.Remember, that economic empowerment and psychological readjustment on the backdrop of the respect of the rule of law by all without exception will form the bedrock of our national prosperity- all prerequisite for true political independence. Note that the form of capitalism that has worked the best for most people most of the time is regulated capitalism-unlike the one we currently know in our country.** Thank you

In September 2006, Mr. Rufus Berry II wrote an article about the benefit of a local company, Liberian Agricultural Company, to the people of Grand Bassa County. This article was not only insulting but with the potential to cause future harm to the people of Grand Bassa and their offsprings. Mr. Berry's article is at www. Theperspective.org. September 2006 edition. Due to his numerous distortions and misstatements, Francis Potter and I, both offsprings of that part of Bassa and first hand witnesses to the events discussed, issued this stern rebuttal. To this very date, we continue to stand by our rebuttal, even if bullets to our breasts. Below is that rebuttal.

REBUTTAL TO MR. RUFUS BERRY'S "GOOD, GOOD,GOOD L.A.C"

Written by FRANCES POTTER , B.SC.(AGRICULTURE) AND LAWRENCE A. ZUMO, MD

Monday, 11 September 2006

ANOTHER SIDE OF THE LAC STORY: REBUTTAL TO MR. RUFUS BERRY'S "GOOD, GOOD,GOOD L.A.C" ASSERTION: *"...Without vision, the people will perish and without courage and inspiration, dreams will die....."-Rosa Parks, 1988*

"Yon chen sensii, yon ni chen zlu-eh. (Bassa Proverb) (You learn inorder to be wise, ...not to become a dummy).

We know in Liberia that we have been caught between the devil and the deep blue sea for a very long time. We, as the real children of the dedicated LAC laborers who sacrificed so much by working long hours and years at LAC (Liberia Agricultural Company, Grand Bassa County) for some of us to the get elementary school education at the Catholic Church run-LAC schools, will not and shall not, remain silent. Hence we will inject our voices at this juncture on a few salient issues.

We know in Liberia that we have been caught between the devil and the deep blue sea for a very long time. We, as the real children of the dedicated LAC laborers who sacrificed so much by working long hours and years at LAC (Liberia Agricultural Company, Grand Bassa County) for some of us to the get elementary school education at the Catholic Church run-LAC schools, will not and shall not, remain silent. Hence we will inject our voices at this juncture on a few salient issues.

Whatever the motives of **Mr. Berry in writing his floral piece about LAC and "its fantastic good to the people of Bassa and the Republic of Liberia"** (see article at The perspective.org, Sept. 5, 2006) will not be addressed here. That will be left between the man, his brain power, and his Maker. Neither will we attempt to assess the biochemical, agronomic, ecological, psychosocial nor geophysical impact that LAC has had on that area of Liberia. That will be the subject of a more rigorous scientific exercise which will be published elsewhere. Any discussions of the financial agreements (if any) between the LAC Catholic nuns who taught us and the company which provided the buildings and books will be left out as well.

It is worthy to note ,though, that the legality of the 1959 LAC contract signed by President William V.S. Tubman and the manner in which this was executed is now increasingly being looked by other interested parties.

To ensure that we are all at the same intellectual wavelength, we will give few firsthand accounts/examples to highlight some of the issues that the public should know about the LAC experience.

Despite the terrible human toll and the long term psychological morbidity associated with indentured servitude and slave labor, some of our compatriots will easily say that there is nothing wrong with that; under the pretext that at least these laborers were provided free lodging and at least one square meal a day. Unfortunately, we do not subscribe to that kind of narrow logic and rationalization.

. Unfortunately, we do not subscribe to that kind of narrow logic and rationalization.

Nobody is challenging the fact that "Liberia needs well-paying jobs in the private sector to enable its reconstruction and the healing process...." We advocate any day for the creation of jobs for Liberians wherever they are. We do need jobs, and all the incentives that should come along with them, including schools, roads, bridges, hospitals etc. However, if the job market has been in someone's backyard for more than 40 years and he could not benefit directly from these incentives, he will think twice before giving his own backyard for the master's next garden.

We would like to ask Mr. Berry if he read or has any knowledge of the contract that created LAC, or has he ever visited the plantation to view the conditions he is advocating so lavishly, or did he just receive his fat check from the LAC management and then ran to the nearest computer to write a beautiful article? If he did not, we would like to suggest to read it. If you do, we would suggest that you reread it-paying keen attention to article II which states, "...the corporation agrees to make an examination of the aforesaid areas (land lying between St. John river and the Cestos river south of compound no. 3, Grand Bassa county and including approximately 300,000 acres; and a section of the land situated along the Tappita-Webbo road and including 300,000 acres. The corporation agrees to make an examination of the aforesaid areas to determine the lands therein which may be suitable for the development of such examination, it shall file with the government, with eighteen months from the effective date of this agreement (March 3, AD 1959) more surveys setting forth the geographical boundaries of said areas as they are determined to be situation for exploitation. The 'concession areas" of LAC shall mean only the area comprised within the surveys...).

We are indeed grateful to have received a good Catholic education in LAC. We must point out, however, that our parents worked diligently ,and many times slavishly, to make this happen for us. Once your parent was fired from LAC, that day was the end of education at any of the Catholic schools in LAC.

One of the authors (F.P.) traveled extensively on the plantation both as a child and for research purposes in 1980's. During the war years, she spent between 1995 and 1997 working in the Bassa counties (Grand Bassa, River Cess, Margibi) as a relief worker. She worked for LAC up to June 2000, when she left for the United States. We do not contend to know everything, but we want you to know that the Bassa people know what they are talking about when they resist any more expansion by LAC for rubber plantation. As an agriculturist by profession, she wants you to know that rubber trees and cereal and other shallow root crops which people depend on for food and other forms do not grow interchangeably. Also as a social being, you ought to know that every man wants somewhere to call his "home" be it a niche, a hut, a village or city.

If you ever visited Yekepa, that beautiful oasis of a city in Nimba County, there was a tiny village called "old Yekepa" where the natives reserved as their own. That village had a few huts, but it was a reminder, in the midst of the beautiful European style city that that was the homeland of the Mano and Gio people.A popular Bassa adage states thus: " a man who has seen the corpse of a boa constrictor does not make his farm in a thicket". The Bassa people have lived around the LAC plantation for almost forty years. Zlor River never got a bridge from LAC. The management always maintained that it was not part of the concession area. Wayzohn or compound # 3 never got a high school nor any token assistance from LAC in the form of support for a neighbor. The road from Buchanan has always been maintained by LAC as a matter of business interest. What's about the road leading from the LAC junction to Barsee Giah's town and beyond, Gen-tro, a double hill about three miles from the district head quarters which was a headache during the rainy season every year as far as we can remember. Frances Potter's mother ,as well hundreds of other women, died due to complications of childbirth along that roadside- just a stone throw from LAC because there was no clinic at the district headquarter. (Besides, the government clinic which had virtually no supplies). We could enumerated hundreds of cases where we were left out but we will leave it at that, hoping you get the point.

The LAC sytem is a system that had severe limitations for the many people at the bottom of the pile(ie. more than 95% of us). Basically, there were no further direct means for education beyond the 6th grade. It was an open secret that you would be offered a job as a tapper immediately upon graduation from the sixth grade. That basically was the end of your road unless you were very stubborn, very lucky to get to Buchanan or Monrovia to hustle further education or were smart enough to get a Catholic scholarship by examination at St. Peter Claver's in Buchanan, Carroll High in Nimba or St. Patrick's High in Monrovia. Many of our classmate never made it out of LAC because of obvious economic and financial reasons. For example, from the 1974 graduating elementary school class of 28-30 persons, only two of us obtained any education beyond the 6th grade.

The better housing and other facilities----when we grew up in Gorzohn, ten to twenty families, a total of 100 persons @ an average of five persons per family shared one communal bath, 3-5 toilet pits and of course one old man cleaned the building twice a day. One of the authors (LAZ) grew up in one of the "better housing" –2 bedrooms with his father, stepmother and ten siblings. FP's were twelve in theirs, and former Liberian Defense Minister Daniel Chea's family may have been six or eight in theirs. There was a creek that ran the length of the camp with about six depots for fetching drinking, washing and bath water. Of course, ring worm, diarrhea, dysentery and other water borne diseases were the norm. There was the "Apollo", a truck converted to school bus to transport students to school and church from the far away camps. If you lived 5 miles away, you had to walk, no matter your age—students from operator's camp and Zeah's camps—about 5 miles away had to walk. The better cars were instructed never to pick up anybody, including children. A driver was fired for giving his children ride to school. In 1978, the Apollo had an accident and more than 3 persons died, two of them were children from the same family—the Toublons, now living in poverty in Buchanan after the man had worked for the plantation for more than 15 years.

The hospital—one hospital to serve the whole plantation! If you lived in division 4 or some of the camps far from the hospital, and the "wound"dresser who served the camp left for the day, you were on your own. You don't know what it means to suffer discrimination from "comers" in your own home- supposed land of liberty. Ask us, we can tell you more. Dr. Zumo's father was not allowed to give his children water from the pump inside the facility he was working in. They had to go through the back door. France Potter got suspended from school for two days because she asked the teacher why they gave the black people water from the pump outside in the sun while they took the white people to a different place in the bungalows.Talking about better facilities, all the camps had no light, of course. We got one light bulb once or twice a month when they provided the"Captain Sinbad" movies. With all due respect, we are pretty sure, Sir, you do not know what it means for a man to be shackled in front of his wife and children. We do. Talking about LAC schools reaching to high school level. We are glad to know that our sweat has paid off. Peace be unto the soul of Kowee Saywrayne (uncle of F.P.) who led the brave struggle in the 1980's for LAC schools to be elevated to at least the junior high school level. F.P. was the secretary then of the LAC Youth Association when we led a delegation to petition President William R. Tolbert in 1978 on this case.

Mr. Berry maybe you do not know the humiliation

of not being able to attend the good school next door because your father does not work for the plantation and you have to live with a friend or relative and assume his surname and tag number,-and of course all the other child abuse you may undergo when you are "staying with your uncle". Please query fellow Liberians, eg. Daniel Chea, Jacob Myers , Isaac Davis. And Mr. Berry, what's about having your dad being the tapper for 10-15 years, toting two heavy, back breaking latex buckets from 4:30 am to 3:30 pm and bringing home ½ bags of rice, 1 gallon of palm oil, 2-5 cups of mackerel and $25/mo. If you don't know how your father smells after those hours of tapping 500 trees, reaping latex, cooked rubber just for you to attend one of the free schools, ask the Tambas, the Flomos and of course, check out the tappers's camp.

Our parents could not strike for better wages, benefits or work conditions even if they wanted to. An attempted strike by our parents in mid 1970's was brutually suppressed by Liberian government soldiers right before our eyes. All we did as children was to cry for our parents. I am sure that never made the national news.

In all fairness, we agree that basic medical and surgical care was "free" for all LAC laborers, employees and their dependents. However, informed consent was not obtained for clinical trials of new medications, medical devices/procedures nor for extraction of live biological/surgical specimens for experimental purposes.

Our central government claimed they could not provide the school or similar services that LAC was providing us, despite the near annual head tax, hut tax, presidential and government officials' birthday and bereavement "taxes", etc that our parents were additionally subjected to.

Uniforms and everything else were the tasks of our parents.We know that our parents were perpetually indebted to Mr. Nadine, the local Lebanese merchant from whom these items were purchased.

Wages, were not living wages in our opinion. One of the author's father(LAZ) earned around $1.80 per day working as a chief cook at the LAC Main MessHall. Whilst his white supervisior, whose job was only to taste the food, was paid presumably nearly twenty times that amount hourly (commensurate with U.S. Labor Regulations and Stipulations) in addition to "inconvenience stipend" that she also received routinely. (Inconvenience, we guess, for being in Africa).

After deductions for additional bags of rice, palm oil, meats, etc take home pay for some workers were as low as a net of 2 cents per pay cycle. Those workers with large families, for mere survival, had to maintain staple food-farms at patches far beyond the perimeters of LAC.

One of the authors (F.P.) today still point out nostalgically and precisely to where her birth village is near Gorzohn camp, LAC which has now been overtaken by rubber trees and "covergrass".

Whilst there were benefits working at LAC, if you do the

math correctly, we believe that our parents contributed more and worked harder than the compensation and benefits they ever received as LAC workers/laborers. Thus it is necessary to put these tangible and intangible factors in true economic and logical perspectives when commenting on the LAC situation.

It would be helpful to know the profit and operating expenses of the LAC company but we are not privy to such information, only the responsible ministries can make that public information public if they choose to. We can only surmise that as a business entity if they were operating at a loss, they would have left long, long time ago.

We are of the opinion that no amount of undying love for black people nor commitment would have kept them in Liberia this long.

There were no sustainable pension plans, disability insurance, workman's compensation, janitor/life insurances, nor severence benefits that we are aware of. LAZ' s father left unceremoniously in 1979 with nothing in hand.

A father of one of us (F.P.) worked for the plantation from the stage of planting the bamboo stilts to mark the places to plant the rubber trees to pumping gas and eventually being promoted to "store boy" at the Cold Storage Commissary all between the years 1960-1978. When he got hit in the chest by a huge watermelon by a customer who had gone to buy his goods and her father was packing them in the booth of his car, lifted his head just in time to get the hit. At the time she was in high school (in Buchanan). When we appealed to Mr. Vavoso, then the general manager, for some compensation to send him to hospital in Buchanan or Monrovia because he was coughing blood, we were told that there was no compensation in the contract with the "laborers" for injury on the job. All he got was a monthly pension of less that $30 up to the time of the civil war. He had to pay for whatever treatment he needed from the package. When he died, his family collected $400 from the LAC management as the total compensation due him for 18 years of service.

His end, that is to say, came as it all started-poor and impoverished, despite his best efforts.

An advise to Mr. Berry would be to let the Bassa people fight their own battle. They have their own children who can now read and write and can understand some of the smooth languages of the politicians and corrupt judges. The Bassa people voted en masse for a government to protect their rights,-among them, the right to live on their own land, be their "own man', making their farms and of course reaping the benefit of their land and labor. There is nothing political or treasonous about a man demanding to stay in his home. This is the 21st century, when many people lost their lives fighting for democracy to come to Liberia. Please leave the decisions with the right government agencies. And of course, we want our say, our home and a piece of the pie—a place to call a home for ourselves, our children and their children's children.

Will the real children of the LAC laborers/workers stand up and be heard !!!!!!!

Hello, Mr. Berry? Where are you?. For your

information, industrial development is the fundamental basis of true national development and a prerequisite for rationally benefiting from globalization (refer to eg. page 70, Manchurian mandate, National Geographic Magazine, September 2006).

If LAC is and has been so good for the people of Bassa and Liberia, why after all these years, LAC has never built a technical school nor a simple agrochemical institute or even a simple rubber processing complex where products (which we are still dying to import) like latex gloves, sneakers, rubber insoles, slippers, rubber bands, car and airplane tires, slippers, condoms, rubber water hoses, "o" rings, rain coats/boats, etc could be readily produced.

Using your logic, Mr. Berry, I am sure you can easily argue and accept that LAC does and did not have to bother with these aspects because these people of Bassa are not capable intellectually and thus would have been a painful waste of company resources. So why does LAC have to bother? Right? Are you still there, Mr. Berry??????

We can only hope that those who can read and write but do not understand what they read and write will not claim any superficial intellectual authority over those who cannot read and write.

We received many favorable response to our rebuttal. A particularly unique response is reproduced below in its entirety for the judging public.

Dear Frances and Lawrence,

I must say that today I saw the article by Mr. Berry and was quite annoyed with his lack of understanding on LAC, Liberia and people in general. Then, fortunately I saw you joint rebuttal and it brought back my own memories, both good and bad of my years at LAC as Technical Manager in 1974.

Firstly, the good part was the Bassa people who were very normal and just trying to make a living themselves and more important their children. The bad part was that myself and the Electrical Engineer, also British , ttried to make reasonable changes and to kerb the corruption being undertaken by the senior expats. In the end they (Mr. Vavoso as security Manager) physically deported the other engineer and tried the same with me but I was able to hide in the bush and arrange for my children to be at Nimba so deportation was not an option for him.

At one time I was able to" break out" and drive to Monrovia to contact the parent company in the USA and then it went on to a major investigation which was not a good time. So after a year I left and went to Bong Mine for a further three years.

I also remember the steep hill on the road behind the plantation which on my own authority was lowered to enable taxis to pass easily to and from Buchanan. The only downside of this was that the small market, with reasonable prices, disappeared as they carried the produce to Buchanan for better prices..... the price of progress.

The corruption side was very serious on the plantation as it was lead by the GM Ken Wild and the Civil Engineer Mr. Schneider plus others and of course all condoned by Vavoso. As you know the original plantation was leased to Pirelli (owned by the Vatican hence the catholic influence) and then sold to Unitroyal of the USA. The price of rubber was really low in 1974 and we were "budgeted" to lose around $2 million but it did not stop the expat bad guys setting up their own logging operation and robbing other aspects on LAC.

When I moved to Bong Mine I took my house servant with me (Borbor Logan) as he was part of the family and when we left Liberia for good we bought him a piece of land in Bassa in the hope that it would substain him and his family.

I have worked all over the world in 33 different countries and I am still working at the age of 67 but Liberia is still special to my wife and our two sons and we follow the news with sadness with the hope that things will get better for the people but this time with much more equality.

Hopefully the five years plans have gone..... "from mats to matresses" and... "self sufficient in rice" and the top politicians in Monrovia have stopped filling their own pockets.

If you wish to contact me or discuss further any of my comments then please do so.

Best regards to both and the Liberian people.

Peter, Ranger and Family

Despite the cessation of internecine hostilities and the institution of a "democratic" government, our children and our parents continued to drop like flies from hunger and disease while member of the old regime returned with a vengeance- bordering on genocide. The cries were louder every day but the band just played on with envoys from democratic countries carting away whatever share of the "elephant meat" called Liberia that they could lay their hands on. A particular midnight call from Liberia was too painful to ignore. An amateur poetic narrative of that call is inscribed below:

@@@@@@@@@@@@@@@@@@@@@@
@@@@@@@@@@@@@

WHY, OLE DEAR, WHY??

Why, dear ole America, why?

A country and its people so ostentatiously generous yet malignantly unforgiving of so many others of the wrong hue;

Requiring others at a stroke of a pen to always forgive and forget;

Cruel, how much more can you be?

Thy steps towards democracy in long pants emulating we try;

But with thy many hands in the dark everywhere

We can only trek to heartbreak, tears and more tears.

In cahoot with your agents and manufacturers our leaders seem to always be;
Merrily economic espionage they seem to commit
With impunity and judicial revenge abounding here-there and everywhere
With democracy kings and queens the other way wryly looking
Is it the same Jehovah God of love and mercy that we all lie before in equal supplication so frequently in churches and temples?

Yes, sand and rain in abundance from where we stand;
With bullets falling from the sky and diseases everywhere;
Unable are we to stand long enough to farm and feed;
Like flies, our children, elders, women and selves continue to weep and drop
Never again to rise upon this earth;

Must we this sand and soil eat as they are the only one so unregulatedly abundant?
So thy will be the only will to be done as you so

very well please

While we continue in this deep delusional slumber

Until the next wake up call, which by then maybe so, so late ?

Or is that in reverse thy moral compass worketh?

Forget ye not, most mighty of masters

The law of consequences unintended remains perpetually the only true fulcrum.

Why, ole dear, why?

Lawrence A. Zumo

(*On behalf of the silenced many who could have written a much better piece)

@@@@@@@@@@@@@@@@@@@@@@@@@@@@@@@@@@

I went into clinical neurology practice after training in New York and New Jersey . It was an exciting time after all those sleepless and tortured nights and days. After managing to deal with the new politics of dealing with doctors and technicians with varying level of training and experiences, post residency period has been very rewarding.

However, the vexing question of Liberia's retrogression despite all the political window dressing remained. I found myself now longer able to remain silent on some pertinent issues affecting that troubled land. The rationale was simple: the life of our children depended on it too and no matter how many years they lived abroad, they had to come to know their homeland too. Thus I began rewearing my double feathered social hat that I wore briefly when I was in high school.

Below are a few selections::

What is a Stroke? How to Recognize, Prevent and Treat Strokes

By Lawrence A. Zumo, MD

February 23, 2006

The recent news stories about Liberia's Archibishop Michael Kpakala Francis' affliction with stroke, as well as reports from Liberia about patients (adults and children) who present themselves to local clinics and health centers with symptoms suggestive of strokes and are often told to go home because this is not a hospital treatable illness and remotely about Israel's Ariel Sharon suffering the devastating consequences of a stroke should be a good point for us to educate ourselves and review some information about this often debilitating and deadly disease. Liberians from as far as the Bujumburam Refugee Camp in Ghana, SouthEast Asia, Grand Bassa County, Lofa County ,Guyana and Jamaica and different areas of the United States have expressed desire to know more about this subject. Hence, I will offer this brief summary.

First and foremost, a stroke is a neurologic emergency. In the absence of a functioning emergency medical service, most preventable strokes will be allowed to exert their often debilitating and deadly effects. Also, health belief modification is very important. The idea that this is caused by witchcraft or God's will is without any basis whatsoever in logic, common sense or science. IN NEUROLOGY, TIME IS BRAIN. Every minute lost in a person with stroke is brain lost, period!!!!.

In general, if someone cannot raise their arms or legs upon command, or cannot speak, you must assume this person is having a stroke unless otherwise proven. It is then extremely urgent to call 911 (if such a system exists) or take the person to the nearest hospital ER or clinic where "stroke teams" and "stroke centers" have been especially formulated to handle such emergencies. If the person is presented to the health facility within three hours with clearly defined onset of symptoms and having met the inclusion criteria, the clot busting drug called tPA can be given and this can be lifesaving for quite a lot of people. This drug is not a cure-all for all strokes but it has made significant impact on strokes and is now recognized as a standard of care in most industrialized nations.

In most countries where stroke services and systems are organized, there has been a significant reduction in the mortality and morbidity from strokes. Hence, investing in local medical and emergency medical services, public education is very crucial. By the time,

Archbishop Michael Francis was put on the plane to Ghana and then to the USA for treatment, the precious "window of opportunity of three hours" has since past and effective treatment then was out of reach for him.

In the United States, for example, stroke remains the third leading cause of death and a major cause of disability. Over 700,000 strokes occur every year (approximately one per minute) and more than 2/3 of these are first time strokes. Had they not invested in emergency medical services, patient education and established dedicated 'stroke teams' and "stroke centers", these figures would have been much, much higher. The financial burden created by stroke is high- no matter which country you live in; direct (eg. hospital and physician care) and indirect costs (lost productivity, family and societal burdens). In the USA, the financial burden was estimated to be $57 billion in 2005 alone.

A stroke occurs when the blood supply to the brain or the neural axis is suddenly interrupted by either a clot in the blood vessel supply that area of the brain or bleeding in the brain from rupture of the blood vessels supplying that territory. What we then see are the effects of the destroyed brain tissues on the structures that they control elsewhere in the body.

Strokes can be divided into two categories: ischemic strokes (80% of all strokes), including embolic and thrombotic strokes; and hemorrhagic strokes (20% of all strokes), such as primary intracerebral hemorrhage(bleeding; as in the case of Israel's Ariel Sharon) and subarachnoid hemorrhage (usually from rupture of outpouching of brain blood vessels).

Moststrokesarepreventable(ie.primaryprevention). The core of primary prevention is reduction of as many modifiable risk factors as possible and evaluation of potentially treatable/lesions, such as lowering high cholesterol in the blood with medications, surgically treating significant symptomatic carotid disease and treating irregular heart beats, especially atrial fibrillation- a very notorious risk factor of stroke.

Strokes can occur in the young child as well as the elderly without preference. The causes and risk factors are however different.

TIA, transient ischemic attack refers to a transient neurologic deficit lasting less than 24 hours with complete return to normal. This however is a herald for an impending stroke-ie. the big one. This is most true within one month of this transient ischemic event. For this very reason, TIA should be considered a medical emergency and immediate testing for preventable causes of stroke is a must. Treating strokes require an active, not a passive approach.

Risk factor modification is important in primary

prevention of stroke. High blood pressure must be diagnosed and effectively treated as is checking and treating elevated blood cholesterol with diet and drug therapy. Also eliminate smoking and heavy alcohol use; engage in a sustained exercise program; detection and treatment of diabetes mellitus, heart disease including atrial fibrillation and daily use of aspirin as secondary prophylaxis. It has been shown in the literature that simple, effective management of high blood pressure (hypertension) alone can reduce stroke incidence by as much as 70% (1).

STROKE RISK FACTORS:

Nonmodifiable:
Age: Advancing age; stroke doubles in each decade after age 55
Gender: More prevalent in men but stroke related case fatality higher in women
Race/Ethnicity: blacks and Hispanics have higher risk/mortality rates than whites
Hereditary: not clear cut but maternal and paternal history may increase risk

Modifiable risks:

Prior TIA/stroke; High blood pressure; Diabetes Mellitus, Cigarette smoking; High cholesterol levels; Heart Disease; Atrial fibrillation; Symptomatic carotid stenosis; sickle cell disease

Potentially modifiable: Heavy Alcohol abuse; Drug abuse (eg. cocaine, heroine); Obesity; Oral contraceptives; Physical inactivity, etc.

EARLY WARNING SIGNS OF STROKE

1. Sudden numbness/weakness of face , arms or legs
2. Sudden confusion or trouble speaking
3. Sudden trouble seeing
4. Sudden trouble walking or dizziness
5. Sudden severe headache (eg. worse headache of your life)

In strokes, prevention is so much better than cure and modification of risk factors remains paramount in the prevention of a first stroke. After a stroke or TIA has occurred, secondary prevention should be started as soon as possible and this should include blood pressure medication, blood thinners (eg. anticoagulants or antiplatelet agents depending on the medical indication), cholesterol lowering medications, carotid endarterectomy/stenting in selected patients for carotid artery stenosis ("narrowing") and continued modification of other know risk factors.

STROKES IN CHILDREN:

Strokes in children differs from that in adults

primarily because of the predominance of congenital and genetic causes. There are also notable differences with regards to incidence, etiology, clinical presentation and clinical course. As is true in adults, disorders of the heart and great vessels are responsible for many strokes in children (2).

The presentation of stroke in children differs from that in adults in the following ways: a) seizures at the onset are more frequent in children whether the stroke in hemorrhagic or ischemic infarction; and b) stroke in the dominant hemisphere produces loss of expressive language, usually as mutism in younger children; fluent aphasia("inability to use or understand spoken or written language") is uncommon in childhood stroke. Residual hemiparesis, epilepsy, mental impairment, and hyperactive behavior are common sequelae. Prognosis is least favorable when there are multiple seizures at the onset of the illness.

In children, cyanotic congenital heart disease is the most frequent cause of ischemic strokes, accounting for 26% . Other common causes of strokes are sickle cell disease, intracranial infection, intracranial hemorrhage, vascular malformations, and occlusive vascular disease(eg. moya-moya disease, fibromuscular dysplasia). Inborn errors of metabolism are a rare cause of stroke in children but not in adults. HIV/

AIDS is becoming an increasingly important cause of stroke in children. Trauma to the neck may predispose to arterial dissection.

Evaluation of patients with stroke:

Triage; ABCs(airway, breathing, circulation,etc); Early consultation with a neurologist or stroke team; Emergency CT scan head or MRI/MRA brain;Thrombolytic therapy (after inclusion criteria are met).

This is not meant to be an exhaustive treatise on stroke but written as a summary for public health education and awareness purposes.

References:1.Gorelick, PB. Stroke prevention. Arch Neurol. 1995; 52: 347-3552.Young, R. Stroke in childhood, Neurology in clinical practice, 2000, eds: WG Bradley et al

A further public stroke education by popular demand.

STROKE: SOME PUBLIC PERCEPTIONS VS. NEUROLOGIC REALITY-2007

The term "stroke" encompasses both ischemic and hemorrhagic disturbances of the cerebral "brain" circulation producing acute or subacute central neurologic deficits. 80% of all strokes are ischemic and 20% are hemorrhagic (meaning bleeding in the brain). Ischemia causes critical hypoperfusion in an area of the brain and/or brain stem.

The terms, mild or major strokes are generally the public's description of disruption of brain function. In the public's eyes, mild stroke is a description of what is obvious to them eg. slurred speech that resolves, slight numbness, transient wobbly walking, etc. Basically, a mild stroke to the lay person means **mild non-motor symptoms and/or non-vital functions disrupting** cerebral event.

A person can have a fairly large **cerebellar** stroke without any motor weakness, speech difficulties or numbness except mild wobbly gait. The public may say this is a mild stroke. This is wrong because this person can in 3-5 days have increase brain pressure, and have a worsened outcome if proper neurologic care is not taken. A good neurologic examiner will pick up dysmetria which most often can be the only finding. A small lacunar stroke in the internal capsule area of the brain can cause complete weakness of both arms and legs on the opposite side with severe disability and you can only see a small (less than 1 cm) infarct on the CT scan of the head or the MRI of the brain;

whereas a large ischemic stroke (as seen on the CT scan or MRI brain) involving the cortical area of the brain can show only mild weakness of the hand and face on the opposite side. The reason for this is how the nerve fibers run down from the brain to its supply body structures and also how the brain is wired by nature.

Depending on the duration and extent of this critical hypoperfusion, neurologic deficits can be either transient (TIA= transient ischemic attack lasting less than 24 hours, RIND= reversible ischemic neurologic deficit) or permanent (completed stroke, infarction = established neurologic damage that is irreversible or only partly reversible).

Due to the disruption of cerebral blood vessels causing localized bleeding and with the potential to lead to increased intracranial pressure, worsening impairment of consciousness and even death, the term mild is not frequently used when "bleeding in the brain" is the cause of stroke.

The more common cause of cerebral ischemia is blockage of the brain arterial supply and/or embolic events (from the heart or distant arteries elsewhere in the body, like from a blood clot in the leg).

Every "stroke" like event should prompt a thorough diagnostic evaluation including a good neurologic clinical examination, 2D Echocardiography, Holter monitor, Bilateral carotid ultrasound, cholesterol, RPR, ESR, homocysteine and blood pressure as well as blood glucose checks, etc to identify the possible cause(s), so that effective measures can be taken to prevent a recurrence which could be more devastating (especially during the first 30 days after that initial, sentinel event).

It is important to point that there are events that may mimick a stroke but are not strokes eg. people under undue stress passing out and then returning to baseline quickly without any residual deficits. These may be syncopal episodes, seizures, initial presentation of multiple sclerosis, nonepileptic cardiogenic events or even psychogenic. Hence a good clinical evaluation by a trained neurologist (not just any medical doctor) is crucial to differentiate and treat these.

REFLECTIONS ON EARLY REPORTS FROM LIBERIA'S TRC-1/27/08

Brilliant dissections of the effects of an event and not the cause of that event is at most only half-brained. Parading victims and narrating their subsequent acts and saying nothing about the planners and perpetrators of these acts is woefully lacking. The news so far from the TRC's deliberations multifacetedly are more painful than soothing. The intended effects from the TRC hearings remain as elusive as ever.

The overriding question to many of us is : "will equal justice be meted out?- whether by the legal process or by rehabilitative means when publicly accused people have been exonerated or so? If the reports during the war years are true, then most of those who carried out commands did them under duress, under the influence of mood altering substances, in self preservation or in self defence. These mitigating circumstances and the question of criminal responsibility then behove us to approach the TRC " statement- taking" very, very carefully or else the whole process will seem like a charade.

If we are to believe that nobody is going to go to jail soon, as we are told, for the TRC-established war atrocities, then as we are reopening these gaping wounds in the search of a solution, then where are the grief and guidance counselors to console remaining relatives of victims when the truth about the death of their loved ones is established? Where are the PTSD (post traumatic stress disorder) counselors for these excombatants who are themselves having difficulties coping to postwar conditions? Where are the sociologists on standby? Where are the telephone hotlines and operators helping people thru this process? Was this not thought about and planned for because again we do not have money for that, too? Or these do not matter so much in Liberia? It is important to remember, as in medicine, if you conduct screening for a disease and are not prepared to do anything about that particular disease, then you have wasted everybody's time. I hope we can take cues from this established fact.

By now we would have hoped that the 4000 plus pages from the CIA's clandestine services (even if heavily dacted) given to the TRC a while back as well as information from other foreign security sources and Liberia's own meager security sources could have given the blueprint for how the TRC would have stacked its deck from the most responsible to least responsible crime perpetrators and proceed in that order, contrary to the current modus operandi. However, it appears that the TRC leadership seems to be operating in a completely different "garden of its own" with its own rules and wishes, releasing information about allegations, real or imagined without regard to the after-effects or irreparable damages to an already afflicted population.

Since most of the fighters were child soldiers or young adults conscripted under mitigating circumstances, a special attention must be paid to CRIMINAL RESPONSIBILITY.

Under criminal law, a committed act that is socially harmful and objectionable can be called a crime if two conditions are true: voluntary act (actus reus) and evil intent (mens rea). Hence it follows that there can not be an evil intent when an offender's mental status is so deficient (eg. mentally retarded), so abnormal (eg. under the influence of mood altering substances) or so diseased as have deprived the offender of the capacity of rational intent. Similarly under eg. duress, the voluntary component of the act must and can be questioned legitimately.

According to the M'Naughten Rule, the 1843 precedent for determing legal responsibility and which has been used widely since, states that the question is not whether the accused knows the difference between right and wrong in general, but it is whether the defendant understood the nature and quality of the act and whether the defendant knew the difference between right and wrong with respect to the act.

Hence those under the heavy influence of psychedelic drugs and extreme duress/cohercion like during the recent Liberian war(s), then the M'Naughten Rule requires us to apply the rule of mitigating circumstances to them as well. Then the question must be: who is ultimately responsible? These petty foot soldiers vs. those who planned and gave these orders? That's why starting with the conflict architects, engineers and generals would have made so much more sense with Liberia's TRC hearings, and not the other way around.

Hallucinogens(eg. LSD, MDMA, designer amphethamines, PCP) were reportedly used widely during the Liberian civil war. Where they came from and who brought them to Liberia is entirely another story. Other anectodal reports suggest that gunpowder mixed with alcohol, etc were also feed to young children before sending them to kill people whom their superiors did not like. We are told most of these kids committed these acts as rites of passage.

Some of the psychological effects seen in people under these kinds of illicit drugs include impaired judgment, illusions, hallucinations, rapid heart beat, tremors, etc. It then should not be a mystery when people under the influence of these mood altering substances say they do not remember what they did or said- ie. momentary insanity.

All of this is to say that for the TRC to be a

genuine TRC, the path of equal justice must be pursued without fear or favor. This means that combatants/ foot soldiers who were under the influence of mood altering drugs, duress or acting involuntarily (regarded as victims as well) must be regarded under the rule of mitigating circumstances and legal, psychosocial services provided to them during and beyond this "truth seeking period".

Importantly, that steps be taken to protect the life, liberty and career(s) of people who are alleged to have committed crimes until they have exonerated themselves and not crucify them by the "court of public opinion". Or else, our TRC would have done a great disservice and harm to all of us.

Finally, since no jail terms are immiment, then grief and PTSD counselors as well as hotline operators should be made available nationwide as soon as possible as these wounds of the war atrocities are opened afresh. If not, serious doubts will linger perpetually about the true intent of Liberia's Truth and Reconciliation Commission and its command staff.

SURREPTITIOUS DRUG ABUSE AND THE NEW LIBERIAN REALITY: AN OVERVIEW

In 1977 as the new 9th grade science reporter for THE SAINTS, the local school newspaper at St.

Pat's in Monrovia, one of my initial assignments was to read up on mental illness and write a report on it. This was welcome news for me since back then as I was interested in everything medical and mental, this was a good opportunity for me to learn as much as I could. To begin my assignment, I sought an audience with Dr. Chief Abua Nwaefuna, psychiatrist and then Director of the Catherine Mills Mental Rehab Center. I met him at his Sinkor 9th Street Office at an arranged time. I introduced myself and said I wanted to interview him about mental illness. I will never forget his intial words. "If you want to learn about mental illness, you have to go to the source. I will not allow you to just interview me like the other fellow from America the other day wanted to do and then you will go back and say you have all the information on mental illness" he said sternly. That set the pace for our encounter and has shaped my journey since then. I followed and tagged him at Catherine Mills Rehab for the rest of that assignment.

I saw many scary, difficult and interesting things to say the least. He showed me many forms of mental illnesses, many which he said were drug induced. Back then, the main drug of abuse in Liberia was marijuana especially the Cannabis sativa variety as well as its legal counterpart, ethyl alcohol (alcohol for short). I also learned that the drugs of choice of the wives of the Liberian elite and well-to-do (especially those who spent any amount of time in America) were amphetamine and barbiturates. Cane Juice-C.J. for the poor people and Johnny Walker Red/Black Label Whisky plus or minus some form of cannibis for the mostly male well-to-do. Interestingly, some of these social secrets have been used successfully against several members of the Liberian leadership over the years during the many contract signing ceremonies on the nation's behalf. The modern day weapons of choice during these many of these signing/negotiating ceremonies seem to be alcohol plus or minus drugs, illicit and licit, laced coffee/tea/food, and free supply of men and women as opposed to gunshot , gunpowder and smoke fish of yesteryears. Yet this important detail seem to be lost in translation all the time. DUI is commonly known as driving under the influence. For Liberians in particular, it is my opinion that it should be known also as " decision under the influence" and this should be scrutinized seriously in this new light (legally, historically and socially) and remedied as deemed fit if we and our children are to have any glimmer of hope for longevity as a nation. Stealing, lying, sexual promiscuity(by both men and women) have been aptly pointed out by observational frontline psychological data as covariables that are very much part of substance abuse. (unpublished data, Janet Zumo, 2007). Imagine what substance abuse/

dependence with its associated cofactors of lying, stealing, sexual promiscuity in addition to pathologic gambling can do in wrecking havoc on the national coiffers as well as in the propagation/perpetuation of corruption in Liberia.

Now to the main gist of this article. Fast forward 30 years to 2007. The emerging picture of drug abuse among Liberians is quite different, troubling and worrying as well. More and more men, women and children are using more hard core drugs chronically. The daily revelations of surreptitious and overt drug abuse by Liberians of all walks of life, women, children and men alike make it imperative for us to familiarize ourselves briefly with some of salient features of the main drugs/substances of abuse, psychological and physiologic manifestations, their systemic effects and how to spot the main signs and symptoms as well as implications for our national life and direction. Additionally, spotting a surreptitious user and nudging him/her to seek real help and counseling could save the life of a mother, father, student, their family members, children, etc or avert a catastrophic national decision that could have lasting negative consequences for all of Liberia. Below will be some salient features of the frequently encountered drugs, illicit and licit. Some addicts cleverly state that the symptoms they have are from environmental/occupational exposure to workplace chemical agents like mercury, lead, thallium, organophosphates, ethylene oxide, methyl bromide, organochlorine pesticides which are neurotoxins but the dysfunctions to the central ,peripheral, neuromuscular apparatus and especially the autonomic nervous system are by and large distinct from the features of dependence, tolerance and withdrawal of substance dependence.(eg. Nouri and Zumo, 1998). Hence neurotoxic disorders will be left out of this overview.

All the substances of abuse have potent acute and chronic effects on the nervous system. The effects of the common legal substances of abuse, alcohol and tobacco, are protean and are well known, so will be only mentioned in passing.

Acute intoxication or overdose of substances of abuse often lead to delirium (confusion with associated autonomichyperactivitylikerapidheartbeat,agitation), stupor, or coma, sometimes associated with seizures, respiratory depression and cardiovascular collapse. Chronic use of most of these agents often leads to drug tolerance or dependence. Abrupt abstinence of a chronically used drug with lead to an acute withdrawal syndrome. Drug abuse may affect the nervous system and human body, indirectly via infectious and embolic consequences of intravenous drug abuse, hypersensitivity or immunologic mechanisms.

On the spot urine drug screening (known as "peeing in a cup" in street parlance) is widely used in the United States for evaluation of persons for possible job termination, legal action or treatment purposes because the medical and social importance of drug abuse is enormous and rightly recognized so in the USA and other developed countries. Actual offending agents can be determined precisely by sensitive yet fairly simple chemical methods such as gas liquid chromatography or mass spectroscopy.

Abuse of opiod analgesics can present in two forms-excessive taking of a prescription opiod analgesic or by street addicts who use the illegal drug, heroin. Heroin crosses the blood brain barrier and its effect on the brain is identical to morphine. When heroin is combined with cocaine, it is called "speedball". Aside from the pain relieving effects, morphine or heroin can acutely produce a sense of rush, accompanied by euphoria. Hallucinations (seeing or hearing things that are not there) may also occur. Signs to look out for are intense skin scratching, small pupils, constipation and urinary retention. Overdose of heroin leads to coma, respiratory suppression, and pinpoint pupils.

Marijuana, Cannabis sativa, with its primary active ingredient of THC, tetrahydrocannabinol, has effect on mood, memory, judgement, and sense of time. Cannabis sativa has more than 400 compounds in addition to the psychoactive substance, THC. Often a sense of relaxation, euphoria and depersonalization occurs. High doses of marijuana produce hallucinations, paranoia, or frank panic reaction. Tolerance does develop with chronic use of marijuana, contrary to popular belief. Abrupt stopping of marijuana smoking causes irritability, restlessness, and insomnia. Marijuana cigarettes are prepared from the leaves and flowering tops of the plant. The usual THC concentration varies between 10 and 40 mg/ per marijuana cigarette but concentrations > 100 mg per cigarette have been detected. Hashih is prepared from concentrated resin of Cannabis sativa. THC is quickly absorbed from the lungs during smoking into the blood and is then rapidly sequestered in the body tissues. Although the effects of acute marijuana intoxication are relatively benign in 'normal users', the drug can precipitate severe emotional disorders in individuals who have antecedent psychotic or neurotic problems. Very potent forms of marijuana (sinsemilla) are now available in many communities, and concurrent use of marijuana (considered a gateway drug) with crack/ cocaine and phencyclidine is increasing at an alarming rate.

As with the abuse of cocaine, opiods, and alcohol, chronic marijuana abusers may lose interest in common socially desirable goals and steadily devote more and more time to drug acquisition and use. However, THC, in and by itself do not cause a specific and unique "amotivational syndrome". Signs to look out for: conjunctival injection (ie. red eyes) and tachycardia (rapid heart beat) are the most frequent immediate physical sings of smoking marijuana. Among regular users, tolerance for marijuana- induced tachycardia develops rapidly. Tolerance to conjunctival injection is the most difficult to develop no matter how long the person has used the drug, and hence the easiest sign to look out for . But mind you, the ever present use of the eye drop, Visine, is used by mask this hard-to get-rid-of physical sign. So if a friend, family member or child is always buying Visine for a supposed chronic eye "infection", become suspicious. Visine is not an eye antibiotic!!!!!

Medical marijuana tablets, however, are sometimes used in very controlled setting for patients with HIV/AIDS, or multiple sclerosis or as antiemetic in chemotherapy but that is the exception and not the rule.

Cocaine, derived from the leaves of the coca

plant, is a stimulant and local anesthetic with potent vascocontrictor properties. The drug can be administered orally, thru the nose, through the vein, or by inhalation after smoking it. Cocaine abuse occurs virtually in all social and economic clases of society. Cocaine produces a brief, dose-related stimulation and enhancement of mood and an increase in cardiac rate and blood pressure.

In addition to generalized seizures, neurologic complications may include headache, ischemic or hemorrhagic stroke or subarachnoid hemorrhage.

Although men and women who abuse cocaine may report that it enhances their libido and sex drive, chronic cocaine use causes significant loss of libido and adversely affect reproductive function; hence the increasing and suprising use of Viagara in this subset of people. Impotence and gynecomastia, persisting for long time even after stopping the drug use, have been reported in male cocaine abusers. Women who abuse cocaine have reported major menstrual cycle dysfunction including galactorrhea, amenorrhea, and infertility. Cocaine abuse by pregnant women, particularly smoking "crack" have been associated with increased fetal congenital malformations, fetal withdrawal syndrome and perinatal maternal cardiac disease and strokes. Long term use of cocaine may cause paranoid ideation and visual as well as auditory hallucinations, a state resembling alcoholic hallucinosis. Alert: But the characteristic alcoholic breath is absent here. Treatment of cocaine overdose is a medical emergency and must be treated in an intensive care unit. An emerging pattern called " crack head" behavior is more and more evident in street addicts, etc addicted to crack/cocaine. These addicts in part exhibit abnormal thinking, significantly impaired judgement and impaired decision making processes that are out of proportion to their physical presentation.

Multiple, concurrent drug use is common among drug abusers. So this fact must be kept in the back of our minds. Drugs have varying effects on different organ systems of the human body. A few examples will illustrate this point.

Drug abuse of almost any form increases the risk of strokes. Drug abuse is the most important risk factor for stroke in people younger than 35 years of age. So if you see a young person in this age group with stroke, don't leave out drug abuse as a possible cause of stroke. Foreign materials injected during intravenous drug use may cause infective endocarditis, which in turn may cause septic embolic leading to stroke, etc. Vasospasm and vasculitis are frequently associated with the psychostimulants like cocaine,amphetamines.

Anoxic brain damage often follows drug overdose, especially heroin and other opiates.

Cocaine is without doubt the most important cause of drug-related stroke and accounts for approximately 50% of all the cases (esp. in age less than 35).

A rare but acute complication of heroin and cocaine abuse is acute myelopathy, ie. spinal cord syndrome, with flaccid weakness of both legs, urinary retention, etc. This may be due to sudden drop in blood pressure exposing very vulnerable portions of the spinal cord as the usually work up including MRI, spinal taps are all normal. Sudden bloody urine, kidney failure and muscle weakness can be seen in those who abuse heroin, cocaine, ecstasy, amphetamine (the erstwhile drug of choice of Liberian female elites) and PCP.

Betty Ford-like center for substance abuse in a confidential setting mid or upcountry away from the hustle and bustle environment of Monrovia could go a long way to help us cope with this menace and tragedy.

According to DSM-IV-TR , the diagnosis of substance dependence takes precedence over substance abuse. Dependence is described as 3 or more episodes in a 12 month period; abuse as 1or more in a 12 month period.. Feeding into dependence is tolerance (marked increase in amount of substance with marked decrease in desired effect); much time/ activity to obtain substance, use and recover from it.. Abuse is characterized by recurrent use which results in failure to fulfill major obligations at work, school or home as well as continued use despite persistent or recurrent social/interpersonal problems, etc.

Substance (eg.ethanol, narcotics, opiods like prescription pain killers) dependence is a medical disease that can be treated most effectively with a combination of pharmacotherapy and psychosocial counseling. Available pharmacotherapeutics, in conjunction with psychosocial counseling suppress withdrawal symptoms and decrease craving so as to improve treatment retention and reduce illicit substance use.

It is important to note that certain patients may be at a greater risk for substance dependence than others; eg. those with a family history of substance abuse; co- morbid psychiatric conditions like depression, mood swings/disorders; chronic medical diseases like hepatitis C, HIV as well as those with resolved pain but with lingering desire for pain medications.

Some questions that you may want to ask: (**DAST-10, Skinner, 1982; Yudko, 2007**); Are you unable to stop using drugs/alcohol even if you want to?;

Do you abuse more than one drug at the time?; Do you feel bad or guilty about your drug use?; Does your spouses/parents ever complain about our drug/alcohol abuse? Have you neglected your family/ children because of you drug use? Have you ever had blackouts or flashbacks ass a result of drugs/alcohol abuse?; Have you engaged in illegal/amoral activities in order to obtain drugs?; Do you feel sick/have withdrawal symptoms when you stop taking drugs/ alcohol?; Have you had medical problems as a result of your drug/alcohol use like memory loss, seizures, headaches, bleeding ulcers, etc?

A higher degree of yes" answers to this questionnaire susuggests harmful behavior and warrants referral for specialized/intensive assessment and treatment.

During discussion with people suspected with substance abuse/dependence, it is helpful to use a nonjudgmental tone and give reassurance that substance abuse/dependence is a medical condition and not a moral failure. You can then suggest referral for counseling and treatment . By so doing you may have helped to save the life and sanity of a mother , father, promising student, significant family breadwinner, or that of a significant political, national or legislative leader.

REFERENCES:

1. So, Y.T. Effects of Drug Abuse on the Nervous system, in Bradley, W.G et al (eds): Neurology in Clinical Practice, Vol. II, 2000, pp. 1521-1527
2. Mendelson, J.H, Mello, N.K., Cocaine and other commonly abused Drugs, in Kasper, Braunwald et al. (eds), Harrison's Internal Medicine Textbook, 2006. 16th ed.ch. 374, pp. 2570-2754. McGraw-Hill.
3. Skinner HA, The drug abuse screening test. Addict Behav. 1982;7:363-371
4. Yudko,E, Lozhkina O, Fouts A: A comprehensive review of the psychometric properties of the Drug Abuse Screening test. J Subst Abuse Treat. 2007;32:189-198.
5. Nouri, S, Zumo Lawrence, Kaufman, H. Autonomic Cardiovascular Reflexes in Chronic Fatigue Syndrome. April 1998. American Academy of Neurology, 50th Anniv. Session Presentation.
6. Zumo, Janet. Unpublished data, 2007. Thanks for many years of immense continued support and collaboration.

IN SEARCH OF A NEW 'UL' PRESIDENT: SOME COMPARATIVE SUGGESTIONS-2008

The search and choice of the new University of Liberia "LU" president should be the least political of all the processes in Liberia. Good diction with an outdated, nonvalidated doctoral degree and excellent political connections and/or manipulatibility are not all to it. In this crucial time of our national existence, good brains, dynamism, vigor, vision and research-worthiness are crucial in a good "LU" president.

Compared with other countries around the globe, the nation's largest center of tertiary learning should have a president (male or female) who will be able to take LU by the bootstraps and by example, motivation and dedication, to quickly take LU to the realm of "respectable universities"-unlike trends in the past.

If this is done, LU also can attract research and development dollars that will be very helpful to Liberia's image and its people and in the same time make LU a center of regional excellence where talented students/ teachers/professors/researchers from the world over can come on sabbatical leave- making this a win-win situation for all.. This is not too farfetched a scenario. We must only be willing to make those necessary sacrifices. We can easily look to Mackerere University in Uganda in pre and post Idi Amin Uganda and see what strides they have made by way of a very big comeback.

The new LU presidency should be an office of intellectual dynamism, inventiveness and curiosity and not one of passive retirement and oblivion. It should be on a rotational basis (5-10 years, etc) and based on results production, not politics!. The person should be chosen based on a closed balloting system among the various deans, search committee members, etc, with the political leadership given a vote in this process but not the singular determinative vote.

The new LU president should not be absolved of the duties of teaching and research. My Gosh!, these are critical functions. How can the selected lady or gentleman remain current among his or her peers at home or abroad, during or after the time of his presidency? This mistake must not be repeated.

As is done elsewhere, the search should be by the president-to-be intellectual peers, for true relevance and not by a remote controlled, archaic system that has served no useful purpose to date.

The selected person should then be able to be surrounded by similarly talented, erudite academic staff. They should be able to put forth a winning formula in short order for the full academic brilliance of LU. In this way, the student body will be challenged academically to produce tangible results along with their teachers. An Honors Collegium can be created easily for supertalented students and teachers. Oh, what a glorious day that will be.

Under the new president's watch, viable students and professors exchange programs with European, American, Asian, South American and Australian universities should be established for much needed cross fertilization of ideas. Efficiency should be demanded

from above, not the other way around. Faculty and deanship on merit system crucially.

Under the new president's watch, an achievable goal is to do the leg work to make LU eg. a US Pell Grant/Student Loan Accepting School, as has been done in many former Communist countries in recent years after the collapse of Communism. There is no reason why Liberia which claims to have the longest relation to the US to achieve this in a short time.

Strides can be made for arrangements to be made for eg. Liberian medical students to be able to rotate at some US, Europe, Australian medical centers inorder to compare themselves on the world scale.

These will truly help the solidify the brilliant legacy of the next LU president, male or female.

CEREBRAL MALARIA: CLINICAL MANIFESTATIONS, NEUROLOGIC AND NEUROPSYCHIATRIC CHALLENGES

By: Lawrence A. Zumo, MD

Malaria eradiction is a task the we must all as Liberians take seriously for if we all join hands, we can untangle the unnecessarily tangled web surrounding malaria eradication in Liberia. Now more than any other time in our history, this end is within reach if we muster sincere, honest willpower and courage.

Malaria and HIV are the twin scrouges still decimating Liberia and the rest of Africa. Malaria worsens HIV disease. New scientific research has proven this conclusively. Doing the current politically expedient thing of focusing on HIV alone is at best one-eyed, at worse brainless._

Malaria has protean systemic manifestations. Below will be a brief review of some of the neurologic manifestations of this deadly but curable disease in attempt to stress the urgent attention that this disease requires.

Cerebral malaria, one of the serious complications of *P. falciparium* malarial infections, is an acute medical emergency. It is an acute, diffuse encephalopathy, with fever occurring in 0.5-1% of cases of Plasmodium falciparum. It is potentially reversible but is fatal in 20-50% of cases, especially among young children (most under age 5) and immunocompromised adults. Unrecognized and untreated, it kills in 72 hours (100% fatality). Hypoglycemia, lactic acidosis, and prolonged coma or seizures are poor prognostic signs.

As is well known and widely reported, world wide prevalence of malaria is estimated to be more than 100 million cases with approximately 1 million deaths per year in Africa alone. One third of all humanity live in malaria infested areas and malaria kills one person, often a child under age 5, every 30 seconds. **WHERE IS THE OUTRAGE??????**

A high index of clinical suspicion, rapid relevant differential diagnosis (eg. meningitis, seizures, toxometabolic etiology,abscess,encep halitis, HIV, HSV encephalitis, hypertensive encephalopathy,exogenous toxins, neoplasms, iatrogenic drug toxicity, nonconvulsive status epilepticus,etc), ready availability of ICU interventions as well as available therapeutics and supplies are of utmost importance.

It must be noted, however, that tissue hypoxia resulting from any malaria infections could pose a risk of cerebral ischemia and infarction in patients with cerebrovascular disease. Neurologic abnormalities observed with cerebral involvement of malaria include: 1) disturbance of consciousness ranging from somnolence to coma, 2)acute organic brain syndromes with altered intellectual function, behavioral changes, or hallucinations, 3) major motor seizures, 4)meningismus (DDx), and 5) focal neurologic signs especially in children such as tremors, myoclonus, and choreiform movements.

Lumbar puncture , done if clinical suspicion exist, is not diagnostic. An increased opening pressure due to cerebral edema and a normal cerebrospinal fluid profile may be present. Rarely, an increased CSF protein and low grade pleocytosis may be observed but hypoglycorrhachia (as in Tb/fungal, bacterial meningitis) does not occur.

It appears that a common basis for these complications includes tissue hypoxia, which results from anemia and alterations in the microcirculation. P. falciparium-infected eryhrocytes undergo major alterations in their rheologic properties that include decreased deformability and adherence to vascular endothelium via specialized RBC membrane modifications ("knobs"). Blood flow through the microvascular beds then becomes sluggish, resulting in decreased oxygen delivery to tissue. Consequently, there is a loss of capillary endothelial integrity and an increase in capillary permeability resulting in leakage of fluid and protein into the interstitial space. If sufficient endothelial damage occurs, hemorrhage may ensue.

The malarial parasite modifies the erythrocyte by exporting proteins into the host cell. One such modification is the expression of PfEMP1 on the host erythrocyte surface which functions as the cytoadherent ligand. The binding of this ligand to receptors on host endothelial cells promotes sequestration and allows the infected erythrocyte to avoid the spleen. Numerous PfEMP1 genes (ie. the *var* gene family) provide the parasite with a means to vary the antigen expressed on the erythrocyte surface. This antigenic variation also correlates with different cytoadherent phenotypes which may affect the parasite's pathogenesis, as has now been demonstrated by the recent discovery of malaria parasite 'cloaking" genes.

Two proteins which might participate in "knob" formation or affect the host erythrocyte submembrane cytoskeleton and indirectly induce knob formation are the knob-associated histidine rich protein (KAHRP) and erythrocyte membrane protein-2 (PfEMP2), also called MESA. Neither KAHRP nor PfEMP2 are exposed on the outer surface of the erythrocyte, but are localized to the cytoplasmic face of the host membrane. Their exact roles in knob formation are not known, but may involve reorganizing the submembrane cytoskeleton . The knobs are believed to play a role in the sequestration of infected erythrocytes since they are points of contact between the infected erythrocyte and vascular endothelial cells and parasite species which express knobs and exhibit the highest levels of sequestration. Additionally, disruption of the KAHRP protein results in loss of knobs and the ability to cytoadhere under flow conditions.

Extreme caution is required in fluid management of patients with complicated falciparium malaria. It is also very important to note that worsening cerebral dysfunction in patients with falciparium malaria, even during therapy, may be a sign of hypoglycemia or gram-negative septicemia This can arise because of eg. increased consumption

by parasites, malabsorption or increased pancreatic secretion of insulin induced by quinine. (Treatment: rapid infusion of 50% dextrose).

Correction of lactic acidosis and anemia, anticonvulsant administration (phenobarbital, diazepam) and intracranial pressure monitoring("bolts",nl ICP <15 mm Hg) are other aspects of patient care. The safety of other anticonvulsants like dilantin, fosphenytoin iv/im and depacon in the setting of severe malaria have not been determined as the data are not readily available from the tropics.

It has however been reported by researchers in Kenya, that treatment of malaria related seizures with Phenobarbital has led to doubling of mortality in children receiving a single prophylactic intramuscular injection of Phenobarbital (20 mg/kg). Unsure whether it was given at the recommended rate of 2 mg/kg/min in children. Reasons for this are not clear: eg. due to respiratory depression (in the absence of readily available respirators), or contributions from other disease-related processes ? Corticosteriods used with quinine prolong coma and worsen outcome, compared with quinine alone, hence not recommended. Exchange transfusion should be considered in patients with a parasitemia level of greater than 10 percent and deteriorating neurologic status.

Concurrent HIV infection has now decisively been related to severity of malaria, and progression or acceleration of HIV infection during malaria co-infection has been confirmed.

In fatal cases of malaria, the cerebral cortical gray matter may contain congested vessels filled with parasitized erythrocytes, perivascular edema, hemorrhage (ring hemorrhage), and rarely glial reaction (malarial granuloma).

As is well known, natural transmission of all *Plasmodium* species (falciparum, vivax, ovale,malariae) causing human malaria is by the female *Anopheles* mosquito. Additional cases are accounted for by laboratory accident, blood transfusion, experimental and zoonotic infection.

The incubation period is influenced by the species of malaria parasite, the degree of acquired immunity, and the dose inoculated. In nonimmune patients, the incubation period for *P. falciparum* is approximately 12 days (range: 6-25 days), and prophylactic drugs may extend the incubation period.

The prodrome of lassitude, myalgia, headache,

chills and rigor may occur before an acute attack. Typical attacks sequentially show shaking chills, fever to 105oF, diaphoresis, with the periodicity of fever determined by the length of the asexual cycle. In P. falciparum, fever is every third day. Several other signs and symptoms do occur. For example, muscle pain and weakness, even myonecrosis, may occur because of increased blood viscosity or capillary obstruction.

Uncomplicatedchloroquine-resistantfalciparum cases are treated with oral quinine sulfate (salt, 10 mg/kg every 8 hours for 7 days) plus pyrimethamine-sulfadoxine or quinine plus tetracycline in regions with know resistance to antifolates. Oral quinine can cause cinchonism with tinnitus, headache, nausea and visual changes (minor toxicity) or rare but major toxicities like cardiac arrhythmias and/or neuromuscular paralysis which can be fatal in patients with unsuspected myasthenia gravis. Mefloquine, an alternative oral agent for treatment and prophylaxis, has been associated with psychotic effects.

The introduction of the new artemisin based combination treatment of malaria offer new hope and promise but the question of availability and cost has to be addressed efficiently and rapidly.

Delayed or late neurologic complications,

designated the *postmalaria neurologic syndromes*(*PMNS*), also occur. The brain is a favored site of parasitized erythrocyte sequestration. In several necropsy series, the finding of parasitized red blood cells blocking brain capillaries are not limited to patients who died of cerebral malaria.

However, structural lesions such as infarcts or hemorrhages are difficult to reconcile with the extent of central nervous system dysfunction in PMNS, and its rapid resolution.An immunologic basis, including the role of TNF-tumor necrosis factor, has been proposed.

Patients who recover from cerebral malaria may relapse within 1-2 days into coma. The CSF contain elevated protein levels (200-300 mg/dl) with or without a lymphocytic pleocytosis. Poor, incomplete, or full recovery occurs. Another, more benign, postmalaria neurologic syndrome follows uncomplicated malaria with *Plasmodium vivax* or mixed infections. Patients are aparasitemic when they develop psychosis, encephalopathy, seizures, tremor, or cerebellar ataxia, which lasts several days to weeks and is usually followed by complete recovery.

In a recent report by Thiam et al, among patients admitted to a psychiatric ward in Senegal, four patients admitted with mental, confusion, delirium, visual hallucinations, motor agitation that were associated with fever, headache , nausea and vomiting were found to be suffering from cerebral malaria. With a high index suspicion, the clinicians were able to make the diagnosis rapidly, treat the patients with neuroleptics in addition to antimalarial agents. Neurolepics were successfully discontinued after 15 days and no relapses were observed after one year of followup.

In a cohort of patients, reported by Carter and Neville, whom they followed at one and nine years after severe falciparium cerebral malaria, 24% (and 10% respectively) of those with cerebral malaria /seizures had greater problems with deficits in neurocognitive domains (executive functions) up to 9 years later.

As they grew older and faced more complex cognitive and linguistic demands, socially and educationally, they continued to face challenges. Their neurologic status at discharge was not a good predictor of later neurocognitive impairment. Hence followup with interaction of educators and therapists is essential but is woefully lacking in the African subcontinent.

In the report by Brewester et al, out of 604 Gambian children admitted with falciparium malaria to one hospital between September and December,

1988, 308 had cerebral malaria and 203 were severely anemic. 14% of those with cerebral malaria died, as did 7.8% of those with severe anemia. 12% of those surviving cerebral malaria had residual neurologic deficit. 69 other children were admitted with clinical features strongly suggestive of cerebral malaria but with negative blood films. 23% of these died and 4.3% had residual neurologic deficits. The commonest sequelae of cerebral malaria were hemiplegia (23 cases), cortical blindness (11), aphasia (9), and ataxia (6). Predisposing factors to sequelae included prolonged coma, protracted convulsions, severe anemia, and a biphasic clinical course characterized by recovery of consciousness followed by recurrent convulsions and coma.

At followup 1-6 months, over 50% made full recovery but about 25% were left with a major residual neurologic deficit. **Hence cerebral malaria may be an important cause of neurologic handicap in the tropics.**

Neurocognitive sequelae are increasingly being recognized as added burden of P. falciparum malaria, in addition to the traditionally recognized burdens of infections and mortality. Due to difficulties with collecting data, the estimation of this burden in subSaharan Africa is only approximate but Victor

Mung'Ala-Odera et al at the Kenya Medical Research Institute estimate that at least 1,300- 7800 children will have neurologic sequelae following cerebral malaria in stable endemic areas per year.

A study in Yemen demonstrated that parasitemic children performed worse than non-parasitemic children in fine motor tasks, but not in cognitive tests. In a randomized control trial of chemoprophylaxis in Sri Lankan school children, chloroquine prophylaxis improved the scholastic performance during a malaria transmission. However it was not clear that this could be equally applicable to African children in different endemicity settings.

Accurate assessment of disability in Africa is difficult due to numerous factors, but forms an integral part of the WHO estimation of disease burden, which is summarized as the disability adjusted life years (DALYs).

Many of the neurologic sequelae are life-long impairments in quality of life and provide a chronic economic and palliative care burden to households. The economic costs of disabilities within households may be overt or through the loss of potential earnings. Many children with severe motor impairments are unlikely to attend school, particularly

when the resources for education are scarce. Thus, Plasmodium falciparum may be one of the most important causes of acquired disability in malaria-endemic areas of the continent (eg. Liberia), leading to long-term personal impairments , in addition to so many other societal ills, that may impede economic development in countries plaqued by the scrouge of malaria.

NEWER DEVELOPMENTS: Researchers from Switzerland have recently begun to use an altered version of heparin (with no effect on systemic anticoagulation) in cerebral malaria to counter the "knob" formation of Plasmodium falciparium-infected RBC membranes.Efficacy and safety studies are pending.

The emergent reuse of DDT and other environmentally friendly insecticides for malaria eradication is a very encouraging sign. But the momentum must be maintained unabated if we are to firmly and finally nail the coffin shut on mosquito, in perpetuity.

REFERENCES

Abu-Raddad ,LJ, Patnaik, P, Kublin, JG, Dual infection with HIV and malaria fuels the spread of both diseases in sub-Saharan Africa, Science. 2006 Dec 8; 314(5805):1603-6.

Bradley, W.G (ed) et al , Neurology in clinical practice, 2000, Parasitic Infections, 1386-1389

Brewster, DB et al. Neurological sequelae of cerebral malaria in children. Lancet 1990; 336: 1039-43

Carter, Neville: J. Neurol. Neurosurg. Psychiatry: 2005, April: 76(4): 467-8

Cowman AF, Kappe,SH, Malaria's stealth shuttle, Science 2006, Sep. 1, 313(579): 1245-6

Crabbs BS, Cooke BM, et al, 1997, Targeted gene disruption shows that knobs enable malaria-infected red cells to cytoadhere under physiological shear stress. Cell 89, 287-296

Deitsch KW, Wellems TE , 1996, Membrane modificationsinerythrocytesparasitizedbyPlasmodium falciparum. Mol Biochem Parasitol 76,1-10.

Gratzer WB, Dluzewski AR, 1993, The red blood

cell and malaria parasite invasion.Semin Hematol. 30, 232-247.

Isselbacher, K, Braunwald, E. (Eds) et al .1994, 13rd edition, Ch. 174, Malaria and Babesiosis, 887-896

Macpherson GG, Warrell MJ, White NJ, Loaresuwan S, Warrell DA. Human cerebral malaria: a quantitative ultrastructural analysis of parasitized erythrocyte sequestration. Am. J. Pathol 1985; 119: 385-401

Mandell, G.L., Douglas, G. (eds), 1985, Principles and Practices of Infectious Diseases, Second Edition, Wiley –Medical, Malaria, ch.232. 1514-1523.

Mitchell GH, Thomas AW, Margos G, Dluzewski AR, Bannister LH, 2004, Apical membrane antigen 1, a major malaria vaccine candidate, mediates the close attachment of invasive merozoites to host red blood cells. Infect. Immun. 72,154-158

Mota MM, Pradel, Vanderberg GP, Hafalla JCR, Frevert U, Nussenzweig RS, Nussenzweig V, Rodriguez A , 2001, Migration of Plasmodium sporozoites through cells before infection. Science, 291, 141-144.

Mung'Ala-Odera, V., Snow, R.W., Newton, C.R.J.C. The burden of neuroconginitve impairment associated with Plasmodium falciparum malaria in Sub-Saharan Africa. Am. J. Trop. Med. Hyg. 2004, 71(2suppl), 64-70

Nguyen THG, Day NP, Ly VC, et al: Post-malaria neurologic sundrome. Lancet 1996; 348: 917-921

Oh, Ss, Voight S, Fisher, D, Yi SJ, LeRoy PH, Derick, LH, Liu SC, Chishti AH, 2000, Plasmodium falciparum erythrocute membrane protein 1 is anchored to spectrin-actin junction and knob-associated histidine-rich protein on the erythrocyte cytoskeleton. Mol Biochem Parasitol 108, 237-247.

Pongponratn E, Riganti M et al., Microvascular sequestration of parasitized erythrocytes in human falciparum malaria: a pathological study. Am.J.Trop. Med. Hyg 1991, 44: 168-75

White NJ. Malaria. In GC Cook (ed), Manson's Tropical Diseases (20th ed). London: Saunders, 1996; 1087-1164.

Zumo, LA, Brannan, T, Greenberg, A, et al, Staphyloccocal Meningitis, www.emedicine.com, June 2007

Commentary: Sexual Deviant Behavior-Paraphilias Leaves Scars on Liberia

08/01/07 - Lawrence A. Zumo, MD, zumoamos@aol.com

The recent revelations about the tragic Eric Goodrige case , I believe, brings us to a stopping point where a few words need to be said about this disorder with reflections on the Liberian society. Much is yet to be said and learned about the case mentioned supra but that will be done in due course by the appropriate authorities.

A few points, however, are in order but specific treatments of this subset of psychiatric disorders will not be addressed here.

The major functions of human sexuality are to assist in bonding, express and enhance love between two people and to procreate (ie. to have offsprings). Paraphilias is defined as unusual fantasies or sexual urges/behaviors that are recurrent and sexually arousing.

Paraphilas are divergent/deviant behaviors in that their participants, for numerous reasons, conceal them and the acts appear to exclude or harm others and to disrupt the potential for bonding between people. Paraphiliac arousal may be transient in some people but more often than not they are lifelong problems with associated interpersonal and civic consequences. Paraphilias seem to be largely male psychiatric conditions.

According to the recent psychiatric classification, there are six such specific disorders: fetishism, fetishistic transvestism, exhibitionism, voyeurism, pedophilia and sado-masochism as well as three residual disorders of sexual preference (eg. zoophilia and necrophilia= sex with animals and sex with dead bodies respectively).

Among the legally identified cases of paraphilias, pedophilia is the most common. Because a child is the object, the act is taken more seriously in the USA (sadly unlike in Liberia), and greater effort is spent tracking down the culprit than in the other paraphilias.

The psychological and physical effects of sexual abuse/ sexual violence can be devastating and long lasting: eg. anxiety, loss of self confidence, mistrust of adults, depression combined with shame , guilt, and a sense of permanent damage, as well as suicidal tendencies, severe marital discord in later life, etc. Abused children tend to be hyperalert to external aggression as demonstrated by their inability to deal with their own aggressive impulses towards others. Posttraumatic stress disorder, multiple personality disorder, high frequency of substance abuse are common in adults who have been abused as children.

Incest is described as sexual relations occurring between close blood relatives. In broader terms, it is described as sexual intercourse between people who are related to each other by formal or informal kinship bond that is culturally regarded as a bar to sexual relations. Eg. sexual intercourse between stepparents and stepchildren or among stepsiblings is considered incest even though no blood relationships exist. Biologic factors strongly support this taboo.

Fathers, stepfathers, uncles and older siblings most commonly sexually abuse children. Mother-son incest is the strongest and almost nearly universal taboo, and this form of incest is rarer than all the others.

Since the news of the Eric Goodridge story, we have conducted an informal survey among Liberian girls and women. Nearly all of them could recall several such incestuous, molestation events involving themselves, their close friends or siblings. When asked why this was never brought up, their near unanimous answer in the usual Liberian way is : "What could we do? We all survived and we are okay. We will leave the rest to God". This sort of resignation is not only bad but dangerous for the psyche of people so afflicted.

When I reflect on our current case load involving these types of patients and see what severe marital discords, worthlessness and depressive symptoms that brought them to our attention in the first place, I am so saddened about the total lack of institutional safeguards and enforceable laws against this sort of heinous crime(ie. pedophilia) in Liberia.

I am usually told that the perpetrators are the usual "saint-by-day and sinners-by night bible totting, pristinely dressed, very religious "bigmen" and "bigwomen" (in the case of Godmas) in Liberia. It makes me wonder whether all their church going is using God as a cheap form of psychotherapy and the common, ordinary people not knowing that this is the case.

One of the recent writers to FPA, Ananys Little urged victims to come out, speaks out and report these to the police. But that is all good in America but how can we get this to be done in Liberia where the need is greatest? I think we have to start with the laws, as well as starting a fast reporting mechanism that can be acted upon there. Some women groups here can give us a starting point because most Liberian men are in denial and starting with them would be like pulling teeth without anesthesia. This is no joke as the psychological scars, in addition to everything else, can be life long.

We have seen several girls take to prostitution to do away with the urge after the initial sexual abuse. We all know that this along with sugar-daddyism, Godma-ism, Godpa-ism do not help us with curbing the spread of HIV/AIDS.

I have traveled the globe over the last two and the half decades. In the company of others, invariably someone who knows about Liberia, will somehow ask: " Are you from that nation of pedophiles (thanks to Pres Tolbert much publicized attempt to lower the legal age of consent) and that country where they have the most ships registered but have never been able to build a single ship? It is a tough pill to swallow but it is what it is. Can we change this image as Liberians? I hope that as we all await the direction of the present case, this would serve as a wakeup call. Will this one ,like most things we do in Liberia ,be pay now or pay more and dearly later?

REFERENCES"

1. Kaplan, HI, Sadock, B.J. Synopsis of Psychiatry, 1998, pp.___ Lippincott,Wilkens

An internet based patient education article published at www.emedicine.com *for the public good.*

Staphylococcal Meningitis

Section 1 of 11

Author: **Lawrence A Zumo, MD,** Neurologist, Private Practice

Lawrence A Zumo is a member of the following medical societies: American Academy of Neurology, American College of Physicians, American Medical Association, and Southern Medical Association

Coauthor(s): **Francisco de Assis Aquino Gondim, MD, MSc, PhD,** Professor Adjunto II, Departments of Physiology and Pharmacology, Neurology Residency Program Director, Faculdade de Medicina, Universidade Federal do Ceará, Brazil; Alan Greenberg, MD, Director, Associate Professor, Department of Internal Medicine, Jersey City Medical Center, Seton Hall University

INTRODUCTION

Section 2 of 11

Background

Meningitis due to *Staphylococcus aureus* accounts for 1-9% of cases of bacterial meningitis and is associated with mortality rates of 14-77%. It usually is associated with neurosurgical interventions (such as cerebrospinal fluid [CSF] shunts), trauma, or underlying conditions such as malignancy, decubitus ulcers, cellulitis, infected intravascular grafts, chronic alcoholism, diabetes mellitus, osteomyelitis, or perirectal abscess. It is uncommon in immunocompetent individuals in the absence of focal infection (eg, pneumonia, osteomyelitis, endocarditis, parameningeal infection, psoas or epidural abscess, sinusitis, tropical pyomyositis), neurosurgical interventions, or congenital dermal sinus. When staphylococcal endocarditis is the source, blood cultures and peripheral and echocardiographic manifestations will point to that etiology.

Pathophysiology

Neonates are colonized by *S aureus* soon after birth; major niches include umbilical stump, perineal area, skin, and gastrointestinal tract. Later in life, major niches include anterior nares, and about 25% of children and adults become carriers. Health professionals; individuals with diabetes receiving insulin injections, hemodialysis, or peritoneal dialysis; patients with dermatologic conditions or HIV infection; intravenous (IV) drug users; and trauma patients have higher carriage rates. Carriers experience more postsurgical infections than noncarriers.

The next step after colonization is penetration through the epithelial or mucosal surface. The mechanisms underlying penetration are not completely understood, but trauma, surgery, immunosuppression, and other infections are predisposing conditions. After penetration and complement activation, *S aureus* is coated by C3b, immunoglobulin G (IgG), or both (opsonization). Staphylococci are then ingested and killed by polymorphonuclear cells and monocytes. Failure of these defense mechanisms can lead to recurrent or chronic infection. Inherited or acquired defects of chemotaxis, opsonization, or polymorphonuclear leukocyte function (eg, due to severe bacterial infections, rheumatoid arthritis, decompensated diabetes mellitus) predispose patients to continuation of the infection process.

Foreign body infection leads to an acquired phagocytic defect. After hours or days of contact with the foreign body, *S aureus* produces a polysaccharide/adhesin substance that causes it to adhere to the foreign body and protects it from the environment. The resident phagocytic population close to the foreign body is not able to kill the invading strain. Anchoring of *S aureus* to foreign substances also modifies its susceptibility to antimicrobial agents. These factors explain the inability of antibiotics alone to eradicate foreign body infection.

S aureus meningitis has 2 different pathogenic mechanisms, as follows:

In the first form, bacteria are introduced during surgery or by trauma or local spreading (especially coagulase-negative staphylococci) from contiguous infection. Bacteria introduced during surgery cause foreign body infection and subsequent postoperative meningitis. Attachment of *S aureus* to foreign surfaces involves interaction with proteins of the extracellular matrix: fibrinogen, fibronectin, laminin, thrombospondin, vitronectin, elastin, bonesialoprotein, and collagen. *S aureus* ligands for these host proteins have been characterized, cloned, and sequenced. Patients with this type of infection have a lower mortality rate than those with hematogenous meningitis, which may be explained by early recognition and less systemic involvement.

In the second group, hematogenous or spontaneous meningitis, *S aureus* is disseminated systemically.

Infection is more often community acquired, and the incidence of positive blood culture results is higher, as is mortality rate. *S aureus* attachment to endothelial cells during septicemia is complex and involves interaction with fibronectin, fibrinogen, and laminin. After adhesion, phagocytosis by endothelial cells and induction of tissue factor procoagulant activity occur. Any localized *S aureus* infection can lead to bacteremia. In the pre-antibiotic era, mortality rate was 82%. Recent studies reported mortality rates between 30% and 40% in non–drug-using patients with *S aureus* septicemia.

Patients with *S aureus* bacteremia can be divided into 2 groups. The first comprises elderly patients with a recognizable primary site of infection and underlying disorders, who usually are already hospitalized when infection starts. Endocarditis and secondary disease foci affect only 10% of such patients, and the relapse rate is lower than in the second group. The second group comprises young patients without identifiable primary infection; they usually have community-acquired bacteremia due to drug use and a high incidence of endocarditis and metastatic foci. The mechanisms responsible for spreading to the meninges are not fully understood. Sustained bacteremia is important but not the sole mechanism responsible for CNS invasion.

The site of CNS invasion during septicemia is still

not clear. It may involve the dural venous system or choroid plexus, where receptors for pathogens have been found. Transcytosis through microvascular endothelial cells is another possible mechanism of meningeal invasion during meningitis. Once bacteria are in the subarachnoid space, host mechanisms are inadequate to control the infection. Meningeal inflammation increases CSF complement concentrations. However, complement concentration is still insufficient and, despite the increased number of leukocytes, opsonic and bactericidal activity are suboptimal, leading to multiplication of bacteria in the CSF.

Once bacteria enter and replicate within the CSF, inflammation of the subarachnoid space ensues because of bacterial (eg, cell wall components) and host factors (eg, prostaglandins, tumor necrosis factor alpha). Alteration of blood-brain barrier permeability leads to cerebral edema and increased intracranial pressure. Meningitis also modifies blood flow throughout the subarachnoid space, resulting in vasculitis and ischemia. Oxygen radicals may contribute to the increased water content, increased intracranial pressure, and changes in blood flow seen in meningitis.

Frequency

United States

In the United States, *S aureus* meningitis accounts for

1-3% of cases of meningitis and is associated with a high mortality rate (about 50% in adults); however, the prognosis for CSF shunt infections is more favorable.

International

Worldwide, *S aureus* meningitis constitutes 0.3-8.8% of all cases of bacterial meningitis. Hospitals with active neurosurgical services generate more cases of staphylococcal meningitis (eg, infection of CSF shunts). *S aureus* is the second most common cause of CSF shunt infections, outnumbered only by *Staphylococcus epidermidis*.

In one study, 38 of 154 (25%) cases of bacterial meningitis during a 7-year period were nonpneumococcal gram-positive coccal infections. The majority of cases were due to *S aureus* and *S epidermidis*. In another study, *S aureus* was present in 21 of 720 (3%) cases of meningitis. Thirteen of the 21 cases were patients in the postoperative period after a neurosurgical procedure, and 3 of the remaining 8 patients had endocarditis or a parameningeal focus of infection.

Mortality/Morbidity

Staphylococcal meningitis is associated with a high mortality rate (about 50% in adults), particularly hematogenous *S aureus* meningitis (mortality rate, 18-

56%). The prognosis for CSF shunt infections is more favorable, probably because of earlier recognition.

Race

Data not available

Sex

Data not available

Age

Newborn nurseries seem to experience waves of staphylococcal epidemics that occur in cycles (ie, epidemics occurred in the 1900s, late 1920s, early 1950s, early 1970s, late 1980s, and early 1990s). *S aureus* was the most common staphylococcal pathogen in the nursery from the 1950s to the 1970s.

CLINICAL

Section 3 of 11

History

With a high index of suspicion, making the diagnosis of bacterial meningitis, in general, is not difficult. All febrile patients with lethargy, headache, or confusion of sudden onset, even if fever is only low grade or the patient is a confused alcoholic, should undergo an urgent lumbar puncture, since a definitive diagnosis of meningitis can be made only by examination of CSF. In patients who have not undergone a neurosurgical procedure, presentation of *S aureus* meningitis may be similar to that of other types of bacterial meningitis. Patients with septicemia have additional systemic signs and symptoms, including septic shock.

- Shunt infections can be insidious, although a fulminant postoperative course can be seen with *S aureus* infection. Coagulase-negative staphylococci (CoNS) are normal inhabitants of the human skin and mucous membranes. Patients most at risk for CoNS infection frequently have a disruption in their host defense mechanisms due to surgery, foreign body placement, or immunosuppression. Because CoNS are common contaminants of cultures, the diagnostic definition of adult CoNS meningitis is different from the meningitis caused by other common pathogens and, hence, is defined with a more strict criteria.
    - Common presentations include low-grade fever in 14-92% of cases, malaise, poor feeding, and irritability. Signs of meningeal irritation are not usually present, since no functional communication exists between

the infected ventricles and CSF spaces in most of the cases.

- Redness of the skin overlying the shunt, if it occurs, is a highly specific sign. Infections with symptoms referable to the distal portion of the shunt are more specific: shunts that end in a vessel lead to bacteremia, while shunts that end in the pleural or peritoneal space cause peritonitis or pleuritis.
- In immunosuppressed patients, the classical meningeal signs may be absent.

- In IV drug users, *S aureus* from bacterial vegetations on cardiac valves is most commonly the starting point for systemic involvement and meningitis.
- In one clinical series, CoNS were reported to be 52.8% of pathogens of ventriculoperitoneal shunt infections in pediatric patients younger than 8 years. Data on adult CoNS meningitis were not given as these had not been specifically examined in the literature.

Physical

- Classic signs include the following:

  - Neck stiffness
  - Altered consciousness (drowsiness, confusion, stupor, coma)
  - Generalized or focal seizures
  - Brudzinski sign (flexion at the hip and knee in response to forced flexion of the neck)
  - Kernig sign (inability to completely extend the legs)
- In *S aureus* septicemia, look for signs of systemic embolization/seeding, which include Roth spots, Janeway lesions, petechiae, subconjunctival hemorrhages, and cardiac murmurs.

Causes

- Hospitals with active neurosurgical services generate more cases of staphylococcal meningitis (eg, infection of CSF shunts) than other clinical facilities.
- In one study, 38 of 154 (25%) cases of bacterial meningitis during a 7-year period were nonpneumococcal gram-positive coccal infections; the majority of cases were due to *S aureus* and *S epidermidis*.

DIFFERENTIALS

Section 4 of 11

Aseptic Meningitis

Haemophilus Meningitis

Tuberculous Meningitis

Viral Encephalitis

Viral Meningitis

Other Problems to be Considered

Behçet disease
Chemical meningitis (eg, after spinal anesthesia, myelography)
Epstein-Barr virus infections
Fungal meningoencephalitis
Legionnaire disease
Leptospiral meningoencephalitis
Listeria monocytogenes meningoencephalitis
Necrotizing cerebral angiitis
Neoplastic angioendotheliosis
Mycoplasmal pneumonia
Rickettsial encephalitides

WORKUP

Section 5 of 11

Lab Studies

- CBC with differential demonstrates polymorphonuclear leukocytosis with left shift.
- CSF analysis is the diagnostic test of choice for suspected meningitis.
  -
    - CSF lactate dehydrogenase (LDH) appears to be diagnostic and has a prognostic value in bacterial meningitis. Increase in total LDH is observed consistently in bacterial meningitis, mostly due to increases in fractions 4 and 5, which are derived from granulocytes. LDH fractions 1 and 2, derived presumably from brain tissue, are elevated only slightly in bacterial meningitis but rise sharply in patients who develop neurologic sequelae.
    - Leukocyte count in the CSF ranges from 250-100,000/μL. Counts above 50,000 raise the possibility of a brain abscess having ruptured into a ventricle. Neutrophils predominate early in infection, but mononuclear cells (lymphocytes, plasma cells, histiocytes) steadily increase as the infection continues.

- Protein content is higher than 45 mg/dL in greater than 90% of cases. In most cases, the protein ranges from 100 to 500 mg/dL.
- Glucose content is usually diminished to below 40 mg/dL or to less than 40% of blood glucose level.
- Gram stain of CSF sediment permits identification of the causative agent in most cases.

- Other laboratory methods for identification of causative organisms include counterimmunoele ctrophoresis (CIE), radioimmunoassay (RIA), latex particle agglutination (LPA), enzyme-linked immunosorbent assay (ELISA), and—most sensitive of all—gene amplification by polymerase chain reaction (PCR).
- Blood cultures should always be obtained. They are positive in 40-60% of patients with *Haemophilus influenzae,* meningococcal, or pneumococcal meningitis, but data are scarce for staphylococcal meningitis. Blood cultures may provide the only definite clue as to the causative agent if CSF cultures are negative and if more sophisticated diagnostic identification procedures are not readily available.

- Because of earlier antibiotic intervention in patients presenting with signs suggestive of bacterial meningitis, a noted rise occurs in culture-negative CSF and blood cultures in some laboratories. This makes the use of a non–culture-based system to detect and identify the causal agents increasingly important. It is here that the 16S rRNA PCR becomes a valuable molecular tool to aid in the detection on nonculturable etiologic agents of meningitis. With the advent of polyacrylamide gel electrophoresis (PAGE) to separate mixed 16S rRNA amplicons prior to sequencing without the need of cloning, the PCR technique is increasingly being used to augment staphylococci identification.

- 16S rRNA genes exist in all bacteria and accumulate mutations at a slow constant rate over time; therefore, they may be used as "molecular clocks." Highly variable portions of the 16S rRNA sequence provide unique signatures to any bacterium and useful information about relationships between them. These properties provide important aids in microbiologic diagnostics, especially in equivocal cases.

- Complement levels and immunoglobulin levels should be part of the evaluation of every patient with bacterial meningitis.

- Antibody levels should be monitored and pneumococcal and meningococcal vaccines should be given to those with recurrent bacterial meningitis because this is common in those with previous head trauma, skull fracture, or dural CSF leak, as well as those with deficiencies of any of the complement components or hypogammaglobulinemia.

Imaging Studies

- Chest x-rays are important because they may show an abscess or pneumonitis, an important consideration for infants and immunocompromised patients.
- Sinus and skull x-rays may show the presence of cranial osteomyelitis, paranasal sinusitis, or mastoiditis.
- CT scans of the head are usually normal but may reveal nonspecific cerebral edema or show previous neurosurgical interventions. CT scans reveal eroding skull lesions and routes for bacterial invasion (eg, mastoiditis, sinusitis, tumors, sinus wall defects, brain abscess, subdural empyema). In patients with immunosuppression or with focal findings, papilledema, or other signs of increased intracranial pressure, a CT scan of the head must be done before the spinal tap to detect

mass lesions that could result in herniation. Those with space-occupying lesions do not undergo lumbar puncture because the withdrawal of CSF removes counterpressure from below, thus increasing the effect of compression from above and exacerbating the brain shift already present. CT scan should be preceded by blood cultures and the initiation of antibiotic therapy.

- MRI with contrast enhancement may demonstrate cortical reactions, including infarctions, hydrocephalus, and meningeal exudates. The role of MRI with contrast T1 and T2 sequences is not well established.
- Transthoracic and transesophageal echocardiograms are helpful for the evaluation of endocarditis. Negative tests do not rule out endocarditis, since neither technique is sensitive enough to detect small vegetations, which may require more than 10 days to develop.

Other Tests

- Lumbar puncture: CSF pressure is elevated consistently (>180 mm H_2O), but pressures greater than 400 mm H_2O suggest the potential for herniation.

Histologic Findings

Pia-arachnoiditis with edema and microinfarcts is observed. Polymorphonuclear leukocytes fill the subarachnoid space in severely affected areas and are found predominantly around the leptomeningeal blood vessels in less severe cases. In fulminant meningitis, the inflammatory cells infiltrate the walls of the leptomeningeal veins and produce a venulitis that can lead to venous occlusion and subsequent hemorrhagic infarction of the underlying brain.

TREATMENT

Section 6 of 11

Medical Care

Bacterial meningitis is a medical emergency. Once purulent meningitis is confirmed by CSF analysis, initial measures include administration of antibiotics with effective CNS penetration and maintenance of adequate blood pressure. Initial antibiotic selection should be based on Gram stain or rapid bacterial antigen tests. If the spinal tap is delayed or the organism cannot be identified rapidly, empiric selection of an antibiotic with effective CNS penetration should be based on age and underlying disease status, since delay in treatment is associated with adverse clinical outcome.

- Standard empirical therapy varies according to age, as follows:

    -

        - In infants younger than 4 weeks, it consists in ampicillin plus cefotaxime or an aminoglycoside.
        - Infants aged 4-12 weeks should be treated with ampicillin plus a third-generation cephalosporin.
        - In children aged 12 weeks to 18 years, a third-generation cephalosporin or ampicillin plus chloramphenicol is an appropriate combination.
        - Adults aged 18-50 years and individuals with basilar skull fracture should be treated with a third-generation cephalosporin, while individuals older than 50 should be treated with ampicillin plus a third-generation cephalosporin.

- Immunocompromised patients should receive the combination of vancomycin, ampicillin, and ceftazidime.

- Patients who have experienced head trauma, have a CSF shunt, or have undergone a neurosurgical procedure should be treated with vancomycin and ceftazidime.

- Vancomycin should be added to empirical regimens when highly penicillin- or cephalosporin-resistant strains of *Streptococcus pneumoniae* are suspected.

- Ampicillin should be added to empirical treatment at any age if *Listeria monocytogenes* is a consideration.

- If allergy to penicillins and cephalosporins preclude their use, chloramphenicol is a reasonable alternative.

- Dose calculations are based on a patient's age and renal and hepatic functions.

- Once *S aureus* meningitis is confirmed and sensitivity determined, therapy may be altered or simplified by using vancomycin, oxacillin, or nafcillin alone. For methicillin-sensitive *S aureus,* nafcillin or oxacillin is standard therapy. If the infective organism is methicillin-resistant *S aureus* (MRSA) or *S epidermidis,* vancomycin is the drug of choice.

- Most experts recommend addition of rifampin

if the patient shows no clinical improvement 72 hours after initial treatment of *S aureus* meningitis.

- Most cases of bacterial meningitis are treated for a period of 10-14 days, except when a parameningeal focus of infection persists (as in most cases of staphylococcal meningitis). In such cases, treatment should be continued for a longer period. Effects of therapy should be tagged to clinical improvement.

- Use of steroids in *S aureus* meningitis is controversial. While adjunctive dexamethasone is beneficial for *H influenzae* type B and pneumococcal meningitis, and some authors favor its use in all types of bacterial meningitis, at present the routine use of dexamethasone is not recommended.

- Shunt removal is often necessary to optimize therapy. If infection is suspected, CSF should be removed from the shunt and sent for studies. Treatment should be started if initial results point to meningeal inflammation and should be modified according to culture results. If infections are difficult to eradicate or if the shunt cannot be removed, direct instillation of the antimicrobial agent is warranted. Daily intraventricular vancomycin doses range from 4-10 mg.

Gentamicin doses are 1-2 mg/day for children and 4-8 mg/day for adults. Combination with an IV agent is always required. Intraventricular teicoplanin also has been employed successfully. Since the entire shunt has a propensity to be contaminated once one section is infected, partial shunt revision is not recommended.

Surgical Care

In cases of *S aureus* meningitis due to septicemia, once the source of infection is identified, surgical debridement or excision may be indicated.

Consultations

Obstructive or normal pressure hydrocephalus may complicate the clinical picture, leading to further obtundation. When either of these is present, neurosurgical consultation for shunting should be considered.

Activity

Bed rest and general supportive measures are needed until the acute illness phase has passed; thereafter, physical activity may be increased gradually as tolerated.

MEDICATION

Section 7 of 11

The goals of pharmacotherapy are to eradicate the infection, reduce morbidity, and prevent complications.

Drug Category: ***Antibiotics***

The agents named are effective in treatment of susceptible bacterial infections such as meningitis due to penicillinase-producing strains of *S aureus.*

Drug Name	Nafcillin (Nafcil, Unipen, Nallpen)

	Interferes with bacterial cell wall synthesis during active multiplication, causing cell death and resultant bactericidal activity against susceptible bacteria; 90% protein bound. Eliminated primarily in bile, 10-30% in urine as unchanged drug; undergoes enterohepatic recycling. Serum concentrations of PO dose peak within 2 h and IM dose within 0.5-1 h.
Adult Dose	500-2000 mg IV q4-6h; 500 mg q4-6h IM for methicillin-sensitive *S aureus*
Pediatric Dose	Neonates (administered IV/IM): <7 days, <2000 g: 25 mg/kg/dose q12h <7 days, >2000 g: 25 mg/kg/dose q8h >7 days, <2000 g: 25 mg/kg/dose q8h >7 days, >2000 g: 25 mg/kg/dose q6h Children: 100-200 mg/kg/d IV/IM divided q4-6h; not to exceed 12 g/d in severe infections

	Documented hypersensitivity
	Associated with warfarin resistance; chloramphenicol may decrease levels; bacteriostatic action of tetracycline derivatives may decrease effects; may decrease effectiveness of oral contraceptives; probenecid may increase levels
Pregnancy	B - Usually safe but benefits must outweigh the risks.
	Avoid extravasation of IV infusions; modify dosage in severe hepatic or renal impairment; elimination rate slow in neonates; caution in patients with cephalosporin hypersensitivity

Drug Name	Vancomycin (Vancocin, Vancoled, Lyphocin)

	Inhibits bacterial cell wall synthesis by blocking glycopeptide polymerization and binding tightly to D-alanyl-D-alanine portion of cell wall precursor. Used in treatment of infections resulting from documented or suspected methicillin-resistant *S aureus* or beta-lactam-resistant, coagulase-negative staphylococci. Also used for serious or life-threatening infections (eg, endocarditis, meningitis) due to documented or suspected staphylococcal or streptococcal infections in patients who are allergic to penicillins and/or cephalosporins.
Adult Dose	15 mg/kg/dose IV q12h
Pediatric Dose	Infants > 1 month and children with staphylococcal CNS infection: 15 mg/kg/dose IV q6h
	Documented hypersensitivity; avoid in patients with severe hearing loss

	Erythema, histaminelike flushing and anaphylactic reactions may occur when administered with anesthetic agents; aminoglycosides may increase risk of nephrotoxicity above that with aminoglycoside monotherapy; may enhance effects of neuromuscular blockade by nondepolarizing muscle relaxants
Pregnancy	C - Safety for use during pregnancy has not been established.
	Caution in renal impairment or those receiving other nephrotoxic or ototoxic drugs; modify dosage in patients with impaired renal function (especially elderly); red man syndrome caused by too rapid IV infusion (ie, dose given over a few minutes) but rarely happens when dose given over 2 h or by PO or IP route; red man syndrome not an allergic reaction

Drug Name	Rifampin (Rifadin, Rimactane)

	Inhibits bacterial RNA synthesis by binding to beta-subunit of DNA-dependent RNA polymerase, blocking RNA transcription. Used in combination with other anti-infectives in staphylococcal infections; management of active tuberculosis; to eliminate meningococci from asymptomatic carriers; and for prophylaxis of *H influenzae* type B infection.
Adult Dose	Synergy for *S aureus* infections: 300-600 PO bid adjunct with other antibiotics
Pediatric Dose	15 mg/kg/d PO divided bid for 5-10 d adjunct with other antibiotics
	Documented hypersensitivity

	Induces microsomal enzymes, which may decrease effects of acetaminophen, oral anticoagulants, barbiturates, benzodiazepines, beta-blockers, chloramphenicol, oral contraceptives, corticosteroids, mexiletine, cyclosporine, digitoxin, disopyramide, estrogens, hydantoins, methadone, clofibrate, quinidine, dapsone, tazobactam, sulfonylureas, theophyllines, tocainide, and digoxin; enalapril may increase blood pressure; concurrent isoniazid may result in higher rate of hepatotoxicity than with either agent alone (discontinue one or both agents if alterations in LFTs occur)
Pregnancy	C - Safety for use during pregnancy has not been established.

	Obtain CBCs and baseline clinical chemistries prior to and throughout therapy; in liver disease, weigh benefits against risk of further liver damage; interruption of therapy and high-dose intermittent therapy are associated with thrombocytopenia that is reversible if therapy is discontinued as soon as purpura occurs; if treatment is continued or resumed after appearance of purpura, cerebral hemorrhage or death may occur

Drug Name	Oxacillin (Bactocill, Prostaphlin)
	Bactericidal antibiotic that inhibits cell wall synthesis. Used in treatment of infections caused by penicillinase-producing staphylococci. May be used to initiate therapy when staphylococcal infection suspected.
Adult Dose	500-1000 mg PO q4-6h 150-200 mg/kg/d IV/IM divided q6h

Pediatric Dose	50-100 mg/kg/d PO divided q6h 150-200 mg/kg/d IV/IM divided q6h; not to exceed 12 g/d
	Documented hypersensitivity
	Decreases effects of contraceptives and tetracycline; disulfiram and probenecid may increase levels; large IV doses increase effect of anticoagulants
Pregnancy	B - Usually safe but benefits must outweigh the risks.
	Caution in impaired renal function

Drug Name	Ceftazidime (Ceptaz, Fortaz, Tazicef)
	Third-generation cephalosporin with broad-spectrum, gram-negative activity; lower efficacy against gram-positive organisms; higher efficacy against resistant organisms. Arrests bacterial growth by binding to penicillin-binding proteins.
Adult Dose	250-500 mg to 2 g IV/IM q8-12h

Pediatric Dose	Neonates: 30 mg/kg IV q12h Infants and children: 30-50 mg/kg/dose IV q8h; not to exceed 6 g/d Adolescents: Administer as in adults
	Documented hypersensitivity
	Nephrotoxicity may increase with aminoglycosides, furosemide, and ethacrynic acid; probenecid may increase levels
Pregnancy	B - Usually safe but benefits must outweigh the risks.
	Adjust dose in renal impairment

Drug Name	Chloramphenicol (Chloromycetin)
	Binds to 50 S bacterial-ribosomal subunits and inhibits bacterial growth by inhibiting protein synthesis. Effective against gram-negative and gram-positive bacteria.
Adult Dose	50-100 mg/kg/d PO/IV divided q6h for 10 d; not to exceed 4 g/d
Pediatric Dose	50-75 mg/kg/d PO/IV divided q6h

	Documented hypersensitivity
	Concurrent barbiturates may decrease chloramphenicol serum levels while barbiturate levels may increase, causing toxicity; sulfonylureas may cause manifestations of hypoglycemia; rifampin may reduce serum levels, presumably through hepatic enzyme induction; may increase effects of anticoagulants; may increase serum hydantoin levels, possibly resulting in toxicity, and chloramphenicol levels may be increased or decreased
Pregnancy	C - Safety for use during pregnancy has not been established.

	Use only for indicated infections, or as prophylaxis for bacterial infections; serious and fatal blood dyscrasias (aplastic anemia, hypoplastic anemia, thrombocytopenia, granulocytopenia) can occur; evaluate baseline and perform periodic blood studies approximately every 2 d while in therapy; discontinue upon appearance of reticulocytopenia, leukopenia, thrombocytopenia, anemia, or findings attributable to chloramphenicol; adjust dose in liver or kidney dysfunction; caution in pregnancy at term or during labor because of potential toxic effects on fetus (gray syndrome)

Drug Name	Ampicillin (Marcillin, Omnipen)
	Bactericidal activity against susceptible organisms. Alternative to amoxicillin when unable to take medication orally.

Adult Dose	250-500 mg PO q6h 500 mg to 1.5 g IM q4-6h 500 mg to 3 g IV q4-6h; not to exceed 12 g/d
Pediatric Dose	50-100 mg/kg/d PO divided q4-6h 100-400 mg/kg/d IM/IV divided q4-6h
	Documented hypersensitivity
	Probenecid and disulfiram elevate levels; allopurinol decreases effects and has additive effects on ampicillin rash; may decrease effects of oral contraceptives
Pregnancy	B - Usually safe but benefits must outweigh the risks.
	Adjust dose in renal failure; evaluate rash and differentiate from hypersensitivity reaction

FOLLOW-UP

Section 8 of 11

Further Outpatient Care

- Monitoring recovery in all aspects is important, including cognitive sequelae, normal pressure hydrocephalus, and seizures.

Complications

- Seizures are more frequent after *H influenzae* meningitis than after *S aureus* meningitis.
- In fulminant meningitis, incidence of strokes is increased because of venulitis, which leads to microinfarcts.

Prognosis

- Untreated bacterial meningitis is usually fatal. A disproportionate number of deaths occur in infants and elderly persons; mortality rate is highest in neonates.
- The presence of bacteremia, coma, seizures, or various underlying diseases (eg, alcoholism, diabetes mellitus, multiple myeloma, head trauma) significantly worsens the prognosis; therefore, an aggressive approach should be used in these settings.

- The likelihood of complete recovery, disability, and employability depends on the underlying condition and the severity of the meningitis.

Patient Education

- For excellent patient education resources, visit eMedicine's Brain and Nervous System Center and Procedures Center. Also, see eMedicine's patient education articles Meningitis in Adults, Meningitis in Children, and Spinal Tap.

MISCELLANEOUS

Section 9 of 11

Medical/Legal Pitfalls

- Low-grade fever, malaise, poor feeding, and irritability in patients with CSF shunts should raise suspicion of meningitis, even when high fever, stiff neck, severe headache, and nausea/vomiting are absent. Failure to consider meningitis may constitute negligence.
- If the patient does not respond to antistaphylococcal antibiotics, appropriate antibiotic coverage must be sought, for instance, by the addition of rifampin.

ACKNOWLEDGMENTS

Section 10 of 11

The authors and editors of eMedicine gratefully acknowledge the contributions of previous author Timothy Brannan, MD to the development and writing of this article.

REFERENCES

Section 11 of 11

- Acar JF, Goldstein FW, Duval J. Use of rifampin for the treatment of serious staphylococcal and gram-negative bacillary infections. *Rev Infect Dis*. Jul-Aug 1983;5 Suppl 3:S502-6.
- Adams RD, Victor M, Ropper A. Principles of Neurology. 6th ed. 1997:695-741.
- Adeloye A. Intracranial suppuration complicating tropical pyomyositis. Report of two cases. *Trans R Soc Trop Med Hyg*. 1982;76(4):463-4.
- Faville RJ, Zaske DE, Kaplan EL, et al. Staphylococcus aureus endocarditis.

Combined therapy with vancomycin and rifampin. *JAMA*. Oct 27 1978;240(18):1963-5.

- Gordon JJ, Harter DH, Phair JP. Meningitis due to Staphylococcus aureus. *Am J Med*. Jun 1985;78(6 Pt 1):965-70.
- Isselbacher, Braunwald, Wilson, et al. Harrison's Principles of Internal Medicine. Vol 2. *13th ed*. 1994:2296.
- Jensen AG, Espersen F, Skinhoj P, et al. Staphylococcus aureus meningitis. A review of 104 nationwide, consecutive cases. *Arch Intern Med*. Aug 23 1993;153(16):1902-8.
- Kilpatrick ME, Girgis NI. Meningitis--a complication of spinal anesthesia. *Anesth Analg* . May 1983;62(5):513-5.
- Mandell GL, Bennett JE, Dolin R. Principles and Practice of Infectious Diseases. 1999:959-997; 2069-2092.
- Millar MR, Keyworth N, Lincoln C, et al. "Methicillin-resistant" Staphylococcus aureus in a regional neonatology unit. *J Hosp Infect*. Sep 1987;10(2):187-97.
- Noel GJ, Kreiswirth BN, Edelson PJ, et

al. Multiple methicillin-resistant Staphylococcus aureus strains as a cause for a single outbreak of severe disease in hospitalized neonates. *Pediatr Infect Dis J*. Mar 1992;11(3):184-8.

- Overturf GD. Indications for the immunological evaluation of patients with meningitis. *Clin Infect Dis*. Jan 15 2003;36(2):189-94.
- Quintiliani R, Cooper BW. Current concepts in the treatment of staphylococcal meningitis. *J Antimicrob Chemother*. Apr 1988;21 Suppl C:107-14.
- Roberts FJ, Smith JA, Wagner KR. Staphylococcus aureus meningitis: 26 years" experience at Vancouver General Hospital. *Can Med Assoc J*. Jun 15 1983;128(12):1418-20.
- Rulison ET. Control of impetigo neonatorum. Advisability of a radical departure in obstetrical care. *JAMA*. 1929;93:903.
- Schlesinger LS, Ross SC, Schaberg DR. Staphylococcus aureus meningitis: a broad-based epidemiologic study. *Medicine (Baltimore)* . Mar 1987;66(2):148-56.
- Schwartz JF, Balentine JD. Recurrent meningitis due to an intracranial epidermoid. *Neurology*. Fe b 1978;28(2):124-9.

- Shinefeld HR. Staphylococcal infections. In: Remington & Klein's Infectious Diseases of the Fetus and Newborn Infant. 4th ed. 1995:1105-1141.
- Spotkov J, Garber SZ, Ruskin J. Staphylococcal meningitis: a complication of psoas abscess. *Neurology*. Jan 1985;35(1):110-1.
- Watanakunakorn C, Tisone JC. Antagonism between nafcillin or oxacillin and rifampin against Staphylococcus aureus. *Antimicrob Agents Chemother*. Nov 1982;22(5):920-2.
- Weinstein MP, LaForce FM, Mangi RJ, Quintiliani R. Non-pneumococcal Gram-positive coccal meningitis related to neurosurgery. *J Neurosurg*. Aug 1977;47(2):236-40..
- Worthington M, Hills J, Tally F, Flynn R. Bacterial meningitis after myelography. *Surg Neurol*. Oct 1980;14(4):318-20.

Realizing a void that had to be filled, at the behest and persistence of my beloved and ever enterprising wife , I became an entrepreneur in real estate investment which has been as rewarding as the practice of neurologic medicine itself.

EPILOGUE

After all those years, the world has become a very complex, tense and uncertain place; making it difficult for the average student to have any real sense of calm that is required for sustained academic excellence. Nevertheless, the conscientitious and diligent student can still find his niche and make his mark in whatever discipline he or she chooses. It will not be easy but it is achievable especially if the path of quiet determination without fanfare is chosen. This synopsis is meant to show what my journey was like. If it enlightens one student, then the effort has been well worth it.

ABOUT THE AUTHOR:

Born in August 1961 to the union of David Kokulo Zumo and Mae Paykue under impoverished conditions in the hinterlands of Grand Bassa County where my father relocated from his hometown of Zolowo, Lofa County to work with the Raymond Railroad Company. I attended St. Joseph's Elementary School , then moved to Monrovia to attend St. Patrick's High School from where I graduated at the top of my class. Tested by the tumultuous events of the April 12, 1980 military coup in Liberia, after several tries, I ended up in Hungary on September 9, 1986 to begin my medical and research sojourn. After my studies in Hungary, I moved on to the USA where I eventually completed my training in neurologist. Additionally, I was expose to the very interesting and equally rewarding world of real estate investing. All these travels covered four continents and spanned more than three decades.

BOOK REVIEWS:

Printed in the United States
141218LV00001BA/61/P

9 781438 923932